GEOGRAPHICAL INFORMATION SCIENCE

NARAYAN PANIGRAHI

CRC Press
Taylor & Francis Group
Boca Raton London New York

CRC Press is an imprint of the
Taylor & Francis Group, an **informa** business

To

my father, Shri Raghu Nath Panigrahi and
mother Mrs Yasoda Panigrahi who brought me up with dedication and placed
education second to none despite their modest means

Table of Contents

Foreword

Over the last decade GIS (geographical information systems) has established itself as a collaborative information processing system. The vast domain of information it can process is ever increasing and so is its popularity. Yet this interdisciplinary field of knowledge is not readily available to the large community of students and academicians as a subject of study.

Traditionally, GIS is looked upon as a processor of geo-spatial information. With the advent of space imaging systems, remote sensing and carto-satellites, the capturing and processing of spatial data temporally has become important. Hence GIS has embarked upon the new role of spatio-temporal data analysis and visualization. Of late, records of various objects can be stored in relational databases and processed using GIS. With the success of GPS (global positioning system), real time tracking of objects and visualizing the dynamics of moving objects on the map has widened the scope of GIS. Therefore, GIS is emerging as a platform for collaborating information from various walks of life having geo-spatial data. GIS gives a unified picture of any operation under consideration.

The author's earnest endeavor to present the subject lucidly and cover the various aspects of GIS from different perspectives is laudable. The IPO (input, processing and output), MVC (model, view and control), and DIKD (data, information, knowledge and decision) models have been used to illustrate the different perspectives.

In my view the author's experience in analysis, design, development and testing of different commercial GIS during his professional career spanning over a decade has been used effectively to compile this text. I believe that this book will kindle the interest of the community of GIS users and readers to further enhance their geo-spatial knowledge.

V. S. Mahalingam
Director
Centre for Artificial Intelligence and Robotics
DRDO, Bangalore

Preface

GIS (geographical information system) has established itself as a popular and effective collaborative information processing system. The input domain of GIS is ever increasing and so is its processing capability and user domain. The scientific concepts behind GIS, known as geographical information science (GISc), are fast emerging as a separate field of study—though an interdisciplinary one.

This book intends to address the GIS user domain encompassing students, users and engineers. Efforts have been made to capture the basics of GIS from the point of view of a student. The requirements of GIS have been explained keeping in mind the general user's level of knowledge. The processing capability of GIS along with the mathematics and formulae involved in arriving at a solution are explained for students and cartographers. The work flow of the whole system, its output and applications are illustrated from an engineer's point of view.

I have tried to present the subject lucidly and cover the various aspects of GIS from different perspectives using the IPO (input, processing and output), MVC (model, view and control), and DIKD (data, information, knowledge and decision) models.

This book may not have had the same impact as it does without the large number of illustrations in the form of figures, tables, block diagrams and references interspersed within the text. These illustrations help in explaining the concepts clearly and are an outcome of a set of trial data prepared particularly to technically evaluate GIS software. Hence in a way they are generic without infringing upon the copyright of any organization or person. I have used many equations, illustrations and references from varied sources and agencies. I have tried to trace the owners of the copyright materials and duly acknowledged the same in the book.

Organization of this book

This book has been organized into fifteen chapters. In order to make the subject simple and lucid to the readers, a systematic approach has been adopted. Each chapter is organized in the form of dialogues, answering the more common questions asked by a reader. Further, each chapter shapes the subject matter to fit the pattern of an 'input–processing–output' model, a 'model–view–control' process, and a 'use case of information' function.

Sometimes the GIS function is compared with traditional cartographic or geographic processes so as to bring out the true advantages and guiding principles of the GIS function. Each chapter is supplemented with work flows and numerous figures illustrating the processes and outputs. Relevant references are added to further the knowledge of the reader. To help in recapitulating the material, pertinent questions are asked at the end of each chapter. Therefore each chapter in the book is organized in the following patterns.
1. The descriptive GIS functions
2. The GIS function as used by the user (use case view)
3. The IPO (input–processing–output) view

4. The MVC (model–view–control) perspective of the GIS function
5. Comparison with the traditional system and procedures
6. Work flow and process flow as applicable in GIS

One more perspective of GIS as an EIS (enterprise information system) has been discussed as the process of evolving EIS. Microscopically each section of the book explains both the IPO and MVC paradigms consistently.

Acknowledgements

This book is an outcome of my association with various GIS intensive projects of the Defence Research and Development Organization (DRDO). These projects involved rich spatial data processing, analysis and visualization functions. The experience gained during analysis, design, development and testing of these projects concerning geo-spatial functions were quite educational and unique. Hence I am very grateful to DRDO as an organization.

I duly acknowledge the constant inputs from my research and development team, which has enhanced the readability of this text. This book would not have been possible without the active help of some enthusiastic individuals.

The academic criticisms of my scientist wife Smita were of great help in structuring the book. My son Sabitra was a source of motivation throughout this project, boosting my enthusiasm every time I felt low. My cute little daughter Mahasweta deserves a mention because I have spent the time, which is rightfully hers, in compiling the book. Subhalaxmi deserves a mention for her witty and critical comments.

My sincere thanks to Mr. V. S. Mahalingam, Scientist-G, Divisional Officer, B Command and Control Division of the Centre for Artificial Intelligence and Robotics (CAIR)—it was he who kindled the very thought of writing this book. His constant guidance and valuable advice were a source of great inspiration.

Finally, I thank Mr. N. Sitaram, Distinguished Scientist and Director, Centre for Artificial Intelligence and Robotics, Bangalore, for his advice and support.

For suggestions/feedback to improve the book you can reach the author at

1

Introduction

This chapter introduces geographical information science (GIS) to a first time reader of the subject. It opens with the first question that would occur to the reader— 'What is GIS?' This is followed with some basic definitions and the genesis of the subject with a historical review explained from various perspectives. In order to generate further enthusiasm in the reader, this chapter is supplemented with workflows, process flow and an information perspective of 'how GIS is evolving'.

1.1 What is GIS?

Geographical information system or geographic information system is an interdisciplinary field of knowledge, comprising geography, digital cartography, computer science, mathematics, image processing, pattern recognition, digital photogrammetry and remote sensing. It is more popularly called GIS.

In another sense, GIS is an information processing system, which is an embodiment of information pertaining to objects on, above and under the earth's surface. GIS primarily processes data which are geographic in nature associated with reference coordinates (latitude, longitude and height). Such data which has reference coordinates associated with it are called spatial data.

1.2 Definition of GIS

To formally define GIS, we must know a little bit about information. Generally, information has three main components—the location (spatial) it is associated with, the time of its happening (temporal) and the contents (what) of the information itself. Any information can be analyzed into its spatial, temporal and non-spatial components by answering fundamental queries such as, where it has occurred? when it has occurred? and what the information is about?

GIS can be defined as a system of systems collecting, processing and analyzing spatio-temporal information regarding earth features. It involves the people preparing the data, the system (input and output devices, computing platforms and networking) and the users using the system.

An alternative definition would be: An information system exploiting spatial, temporal and non-spatial data pertaining to terrain objects. Hence, GIS is an information system capable of answering the following queries:

1. Given the coordinates of an object, what is the object about?
2. Given the description of the object, where is the object located on the earth?
3. Given the location and characteristics, where and when does the object occur on the Earth (time of occurrence)?

Collating all these statements, we can define GIS as an information processing system, collaborating data and information from various walks of life, giving a unified visual representation of the process under consideration with a digital map in the background. Depicted in Fig. 1.1 is an ortho-photo with typical earth features like roads, buildings, trees etc. in it. Also attached to these features are the spatial and non-spatial data corresponding to them.

Road junction at:	Lat, Long, Alt: 12:58:21,77:38:11,846
Road names:	Ring Road and Peripheral Road
State:	Northern Province
Road width:	100 meters
Bi-lane:	Y
All weather:	Y
Date of survey:	07-06-1996

Fig. 1.1 Examples of spatial, non-spatial and temporal data

The data record in Fig. 1.1 constitutes three parts. The first field of the record gives a series of coordinates in the form of latitude, longitude and altitude, describing the physical location of the object and is called the spatial information, because it is related to the object in space. The rest of the data fields barring the last field describe various attributes of the object and are known as non-spatial information or aspatial information. The last field of the record gives us the date and time at which the information is recorded or surveyed. This gives the temporal dimension to the geo-spatial information.

The corresponding visual representation of this geo-spatial information can be found in the digital ortho-photo. This is captured in the form of a digital line feature in a digital vector map and finally as a line with particular colour, line style, font and width in a paper map as depicted in Fig. 1.1.

In this rapidly expanding information age, a plethora of information systems are emerging. The sole driving force behind this information explosion is 'How easily can we provide processed information to the user, so that he or she can make effective decisions'. Most information systems are oriented and developed to address the contents of the information in various ways pertaining to a particular domain. But of late, there is a growing demand to address the spatial and temporal occurrence of the information along with the information itself. Some operation information systems OIS have more need for real-time spatial information, for example, the fleet management system; the tactical command, control and communication system; and the battle field management system etc. Whereas, in some operational information systems such as the urban planner, agriculture and land management systems etc. the spatio-temporal data collected over a period of time is required to be processed. These systems do not exploit the time criticality associated with the spatial data. GIS has emerged as a reliable provider of this kind of spatial information, which need to be processed (Longley et al. 1999, pp. 8–13).

1.3 Genesis and Historical Perspective of GIS

GIS emerged due to a number of operational necessities arising from the usage of digital maps in civil and military applications. Some prominent applications, which also had a profound impact on its evolution, are operation planning, situation representation and terrain feature measurement. The motif force which spurred the development of GIS as an information system was thematic cartography i.e. composition of maps according to a particular theme, collaborative visualization of operation information and application-specific map generation e.g. communication map, soil map etc.

Cartography is the art and science of map-making. It involves surveying objects or features on, above and under the earth's surface e.g. land survey, aerial survey and survey of coastal zones etc. Traditionally, the surveyed features were then represented with appropriate graphic symbols in a paper format such as maps. The graphic symbols representing the surveyed features with specific fonts, colour and styles were collectively called cartographic symbology. The process of surveying and then engaging in the painstaking task of drawing the symbols was human-intensive and carried out by a specialist team of professionals called field surveyors and cartographers. As the need of application-specific maps increased, there was a growing demand for a flexible library of digital cartographic symbols (symbology) for representation of spatial features and a process to prepare application-specific maps. The characteristics of the symbol library are listed in Table 1.1.

Table 1.1 Characteristic requirements of a symbol library

Characteristic	Cartographic requirement of symbol library
Uniqueness	Unique and unambiguous representation of surveyed data in different formats e.g. paper map, sand model, globe or digital graphic display etc. This is achieved through the assignment of a unique combination of colour, line style, font and width or thickness to each set of digitized features.
Scale independence	The symbols can be scaled up or down so as to preserve the relative position of the features on the terrain. That is, the line and area symbols should get appropriately scaled up or down to fill the corresponding space in the map, whereas the point symbols should maintain the location accuracy of the features they represent.
Updatable	The symbology library can be updated from time to time to incorporate new spatial features. To take care of new natural or man-made features, cartographers modify and update the symbol library from time to time.
Temporal change	Random and periodic change in terrain surface is inevitable. The symbol library needs to cater and represent temporal changes effectively. It also needs to be updated from time to time to incorporate such temporal changes or new spatial features.
Application-specific symbol library	The symbology can represent features specific to a particular application such as defense application, civil application, cadastral application etc.
Overlay representation	Man-made features or the operations carried out by an organization need to be depicted on a map. The symbol library should effectively represent the operations without affecting the background map.

As mentioned before, the traditional cartographic processes were quite human-intensive, inflexible and error-prone. With the advent of digital processing technology, the traditional cartographic processes were replaced by digital cartographic processes, which were simple and less error-prone. When remote sensing and satellite photography developed, the manual survey data were supplemented by satellite images which could reflect recent terrain changes and thus were more authentic. The actual survey itself was augmented by a combination of digital cartographic processes such as, survey using GPS (global positioning system), aerial photography and photogrammetric interpretations of remotely sensed satellite images and aerial photographs for accurate representation of earth features.

The whole idea of graduating from the manual cartographic process to the digital cartographic process was to reduce the dependency on human skill, thus removing the errors caused due to this; and accurately representing spatial information in a composite manner.

Earth features are digitized from the satellite image and after due interpretation, are assigned an appropriate symbol, colour, font and style (symbology) to give a visual representation to the feature. Further, these features are categorized according to the common major and minor characteristic they possess (theme-wise). The categories of symbols along with their spatio-temporal information are stored in various data organization methods (formats) so that they can be modified and retrieved from the digital database with ease. The process flow in Fig. 1.2 describes the various stages of thematic cartography. The various modules of thematic cartography are realized in the form of software, hardware or a combination of both. Geographical information systems has emerged as an embodiment of all these digital cartographic modules.

Digital cartography emerged during the 1960s. Many basic concepts in digital cartography (spatial data, map layers, topological structure etc.) can be traced back to work done in the land inventory branch of the Canadian Government or the Harvard Laboratory for Computer Graphics and Spatial Analysis. Some of the first GIS systems for mapping and spatial analysis were put into operation by the Environment Systems Research Institute Inc. (ESRI). ESRI launched its commercial GIS ARC/INFO® at the same time Intergraph's Geomedia® suit of solutions came to be considered as the first generation GIS. These popular commercial GIS systems (Goodchild et al. 2001) owe their origin to Harvard laboratory technology.

GIS systems were first used in the armed forces because there was a felt need for an operation information system (OIS) which could depict the deployment of forces on a thematic map as an information layer. This kind of a system would help commanders to assess the deployment of their own forces with respect to the enemy deployment.

1.4 Process Flow of GIS

The block diagram shown in Fig. 1.2 depicts various processes generally followed in a digital cartographic system. Almost all GIS systems perform some or all of these steps in different ways. The block diagram clearly describes the various stages that geo-spatial data undergoes in a typical GIS system before it can be used as geo-spatial information by the end user.

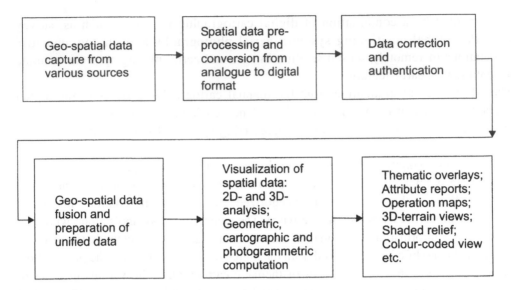

Fig. 1.2 Different stages of processing geo-spatial data

STEP 1   Spatial data can be obtained from various sources e.g. manual survey, scanned paper maps or photos, survey data obtained from global positioning systems (GPS), satellite images, ortho-photos obtained by air-borne surveillance platforms etc.

STEP 2   Some of the spatial data obtained in STEP 1 are analog data and need to be converted to digital form by digitization. Some data do not have geo-reference information i.e. the information required to refer them to a particular patch of earth. Hence STEP 2 involves pre-processing the spatial data obtained in STEP 1 i.e. converting them to digital form, imparting uniform coordinate information and register to the earth surface it belongs to.

STEP 3   The data obtained after pre-processing sometimes has some spatial error due to registration. This needs correction. The spatial data also needs to be validated using ground truthing of a candidate sample before analysis.

STEP 4   The geo-coded, geo-referenced spatial data, obtained after validation and verification through the above steps represents different aspects of terrain viz. two-dimension spatial features, three-dimension spatial features and attributes of different features. Hence to provide a comprehensive view to the user, the spatial data need to be fused for a unified representation of the terrain. This is not a mandatory step in all GIS.

STEP 5   The spatial data obtained in the above steps can be visualized and analyzed in various ways e.g. 2D-digital map, 3D-terrain model etc. Geometric and cartographic computations to measure the spatial features and derive further meaning can be performed on these data.

STEP 6   By applying different processing tools and algorithms provided in GIS, the user can derive many specific spatial information and outputs.

1.5 Evolution of GIS

Because of the collaborative nature of GIS, information from various walks of life e.g. transportation, mining, town planning, civil engineering, water management, operation planning etc. can be processed and viewed together with respect to a map as the background information. Data from various sources are collaborated as spatial and temporal data that can be viewed with a map or 3D-model of the earth as the background. This gives a true picture of the operation, be it mining or water management as deployed on the ground. A unified view helps convey the operation information more precisely to the planner, decision makers, and executors, with respect to space and time.

Evolution of GIS can be traced from its infant days (Longley et al. 1999, pp 8–11) as an application for depiction of operation information in the form of red and blue forces over a digital thematic map. In this section, a systematic approach has been adopted to study the advance and evolution in GIS by superimposing various impact parameters over the trends in the geo-spatial process. Different perspectives and aspects of GIS have been discussed with respect to its contributing technologies. The significant impact parameters of a spatial information system are discussed under the following heads:

1. Geo-spatial data or input domain
2. Spatial data processing techniques
3. Spatial data fusion and visualization techniques
4. Information architecture

1.5.1 Input Domain

With the rapid advance in satellite technology, the frequency, quality and resolution of spatial data acquisition has undergone a sea change. This change in technology has obvious impact on the quantity, quality, and cost of geo-spatial data available for processing. High resolution images of the earth's surface in the range of meters and sub-meters are being obtained by various specialized carto-satellites. The frequency of obtaining the image of a particular patch of earth has increased, enabling GIS to carry out temporal processing such as terrain change detection and feature extraction by comparing two satellite images of the same area taken at two different times.

The broad reach of GIS to various spatial information users and its capability to collaborate with different types of data is an indicator of its increasing input domain. As the input domain of GIS increases, so does its application areas. Increasing input domain means a greater understanding of various spatial data formats and the operations that can be applied on these types of spatial data. New ways of spatial data collaboration and exchange are continually evolving. This drives GIS technology to become truly an inter-operable information system.

1.5.2 Processing Techniques

The processing capability of GIS must keep pace with the growing geo-spatial input domain, application domain and user domain. Earlier the processing capability of GIS was restricted to applying different coordinate transformations, map projections and datum transformations to a set of spatial data for reading location information accurately. Techniques of measurement of distance, perimeter, area, volume, slope, aspect etc. were restricted to a few traditional geometric methods. Of late, the geo-processing capabilities of GIS systems have been expanding in many ways. Geo-processing engines are now capable of processing multi-coordinate, multi-scale and multi-dimension data by applying various soft computing techniques. There are many variants of measurement functions capable of extracting information even from degenerate spatial data. Robust computational geometric algorithms have been set up to handle degenerate spatial data and extract information from them. Traditional data structures and computational geometric algorithms have been modified to handle large volume of geo-spatial data. Soft computing techniques such as artificial intelligence, neural network, pattern recognition and image processing techniques have been augmented with different algorithms to handle large volume of spatio-temporal data. Spatial decision support systems are evolving to assist the decision makers in making use of geo-spatial information. Thus, with the evolving processing capabilities in terms of better data structure and robust algorithms, GIS is fast emerging as a platform for applications which need components such as decision support system, simulation, scientific visualization etc.

1.5.3 Visualization Techniques

Geo-spatial data are characterized by their sources and the agencies that they are obtained from, by their scale or resolution, their dimension and coordinate system etc. Generally geo-spatial data are multi-dimensional (2D, 3D etc), multi-scale (1:25K, 1:50K etc), multi-spectral (infrared, ultraviolet etc.) and multi-sensor (IRS, IKNOS etc). In addition to the above characteristics, spatial data is characterized by the time of its origination making it time dependent or temporal data. All these facets of spatio-temporal data are being utilized by GIS to help in visualization of the terrain under consideration. Thus, GIS has emerged as a base system for multi-sensor, multi-dimension spatial data fusion. Of late, GIS has been giving many views of a patch of earth such as a 2D-view as a map, 3D-views from different observer positions, an orthomorphic view from space and sun-shaded relief and colour-coded elevation views. GIS as a simulation platform for a fly-through or walk-through visualization over the earth surface has many scientific and tactical applications. GIS is emerging as a visualization tool for virtual reality and scientific computation involving spatio-temporal data.

1.5.4 Information Architecture

Geo-spatial data has economic as well as strategic importance. Therefore, the data needs to be secured and collaborated among a strategically interested group of users. This calls for making spatial data available to various authentic users through a local area network (LAN) or a wide area network (WAN). Making the data available through a LAN/WAN network in turn calls for imposing security classification—therefore a mechanism is needed to share spatial data across a secured LAN/WAN. GIS has evolved to handle this twin problem in a unique way. To protect the spatial data from unauthorized users, techniques such as encryption of spatial data, encryption of the header information of the spatial data, encryption of the device storing the data have been adopted. To protect the spatial data in transit over the LAN/WAN network, security measures such as internet protocol security (IP Sec), secure socket layer (SSL), transport layer security (TLS) measures are being embedded into GIS processes. The evolution of information architecture from desktop GIS to an *n*-tier information system of GIS can be attributed to growth in computing science, digital communication and networking technologies.

GIS has evolved from an isolated island of information to an inter-operable collaborative information system. This means evolution of GIS architecture from a stand-alone information system to a client–server architecture, a three-tier architecture in particular and *n*-tier architecture in general. GIS has manifested itself as a true distributed information system adapting to diverse input domain and operational requirements. Also GIS has been ported to diverse hardware platforms such as servers, hand-held navigation devices and high-end display devices. This aspect of GIS will be explained in the last section of this book. The evolution of GIS architecture is explained below.

Earlier GIS was manifested as a single system catering to a single user. This type of system is known as a desktop GIS system. Only one user has the privilege of using all the geo-spatial data and the tools stored in the system at a time. The disadvantage is that other intended users of the system are left starving until the current user frees the system resources. Because of the need for sharing spatial data, GIS has evolved to a form of client–server architecture whereby spatial data is shared by a group of users. In a client–server architecture, the spatial data and the programme serving the spatial information resides in the server. The server essentially receives and responds to user requests by serving the required spatial information. The client–server GIS was the first step towards a distributed spatial information system. The clients are connected through a network of computers. It is a heterogeneous database management system powered by a rich set of geo-processing and visualization tools giving information regarding where and what information of any object on earth. Of late, GIS has emerged as a truly distributed information system by distributing geo-spatial information through the World Wide Web (WWW). In other words, GIS has graduated to an enterprise information system deployed in *n*-tier architecture over a wide area network (Longley et al. 1999, vol. 1, pp. 307–345).

The advancement in GIS is also proportional to technological advancement in hardware, software, and peripheral technology with high-end data collection systems such as GPS, surveillance satellites, unmanned aerial vehicles (UAV) etc. With the advancement in technology, the input data domain and the cycle of capturing the data have also increased rapidly. Basically, GIS is an application-driven technology. Applications in turn are user-driven and every application has some scientific basis in it. With so much happening in the field of science, it can be understood why the methods to process, display and analyze spatial data are still nascent and emerging.

From the foregoing discussions it can be concluded that 'GIS is a toolbox for handling geo-spatial database'. The 'S' in the GIS can be connoted as 'science', 'system', 'service' or 'subject' depending on the user group and the usage. GIS has different methods for different problems concerning different spatial input and user domain.

A decade back GIS was considered as a set of tools for processing, analysis and visualization of geo-spatial data. Of late, GIS has graduated to a full-fledged information system capable of integrating different hardware and software systems. This aspect of GIS as a system will be explained fully in Chapter 13, 'Application of GIS'. Complex functions such as spatial data processing, analysis and visualization have clearly established the scientific worth of GIS. Many paradigms of geographic information science are clearly evident from the growing number of literature being published in the form of research topics, journals, books and websites (List of journals concerning GIS, important websites giving online resources regarding GIS are enlisted in Appendix E.).

Of late, GIS has significantly progressed in many application areas germinating many practical problems and research topics. Therefore GIS is being established as a topic of information science. Professor Goodchild first coined the term 'geographical information science' in 1992 in one of his research papers (Goodchild 1992). The scientific aspects of GIS have been clearly established in Chapters 3, 4, 5, 6 and 7 of this book. These chapters will give the academic and student communities a chance to delve into the scientific basis of the subject.

1.6 Work Flow of GIS

An information system can be better understood by observing the way the user is going to interact with the system for retrieving information. This is generally depicted in the form of a work flow, information flow or data flow etc. The block diagram given in Fig.1.3 depicts the complete work flow of a GIS system. User interfaces (UI), or graphic user interfaces (GUI), which acts as a man–machine interface to a user plays an important role in realizing the work flows of a GIS. A well designed GIS will have a simple and intuitive GUI with lots of online help to guide the user in realizing the GIS functions.

In the subsequent chapters, each of these blocks will be explored functionally and their roles extensively discussed. To make the subject more complete and scientific, a special chapter on the mathematical basics of the process in GIS is dealt with in Chapter 4.

During the past decade, GIS had been considered only as an application software for capturing, storing and analyzing spatial data. Of late, the capability of GIS to model spatio-temporal phenomenon has made it a favourite platform for scientific visualization involving spatial and temporal data. Currently GIS is being extensively used as a decision support system aiding decision making processes in many walks of life. The application areas of GIS are expanding day-by-day and so is its popularity.

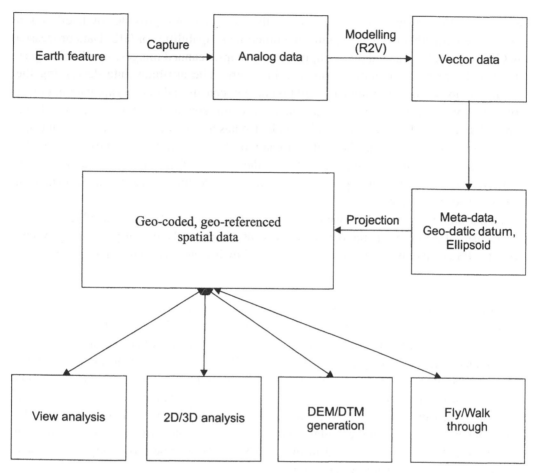

Fig. 1.3 Work flow of GIS

As an information system, GIS can be treated as a processor of spatio-temporal data; a set of tools allowing the user to perform various 2D and 3D geometric processing; a set of tools to perform spatial queries on spatio-temporal data etc. The output of the GIS can be a visual result of the phenomenon under consideration with a map in the background. Various thematic map layers, reports and attribute tables resulting from even one spatial query gives a multiple view of the spatio-temporal information.

1.7 GIS and the Emerging Information Architecture

GIS is an interdisciplinary field of knowledge deriving its application and information domain from geography, cartography, computer science, photogrammetry, computational geometry and mathematics to name a few. The capability of GIS can best be explained by analyzing the input domain it processes, the sources of input data, the applications it supports and the systems it can interface with. The function it provides by itself and in conjecture with other devices explains the functional capabilities of GIS. Data processed by GIS is typically scanned maps, digital vector maps, scanned images, satellite imagery etc. which are predominantly geometric in nature. The attribute data describing the geometric objects are both numeric and textual e.g. census and demography data, power consumption data, per capita income data etc are non-spatial in nature. The spatial and aspatial data are collected periodically and hence has a time factor associated with it i.e. temporal nature of the data. The spatial domain of data (varying with time) exploited by GIS ranges from data beneath the earth, on the earth surface and in space. Thus GIS exploits both temporal and spatial information beside the conventional information attributed to earth objects.

Spatial data are obtained by various means e.g. scanner, GPS, UAV, satellite etc. One mechanism of capturing spatial data at any instant of time is the global positioning system (GPS). GPS is the only system today, capable of showing the exact position of an object on the earth anytime, anywhere, in any weather condition. A constellation of GPS satellites, 24 in all, orbits at 11,000 nautical miles above the earth surface. Ground stations located worldwide continuously monitor them. The satellites transmit signals that can be detected by anyone with a GPS receiver. Using the receiver, one can determine the location with great precision. That is why GPS is an important input device to collect, collate and update spatial data. Therefore, integration of GIS with GPS is an all important application area, which is emerging day-by-day making GIS–GPS a formidable combination for design and development of many creative products. GPS provides the location, velocity, acceleration and time information pertaining to a stationary or moving object, while GIS provides the platform for visualization of the digital terrain information in the form of a 2D- or 3D-digital map. GIS–GPS combines are used in the design and development of many navigational systems for vehicles in land, water and air.

Because of the versatile information domain processed by GIS which is capable of answering both spatial and temporal queries, GIS finds wider application in many fields of engineering such as civil, mechanical, avionics, and telecommunication to name a few. It is also used extensively as an operational information system (OIS) in the armed forces.

Emerging computer network and its architecture has its fallout in graduating GIS from being an isolated information island to a desktop GIS (Fig.1.4 (a)) to an enterprise information system (EIS), making it truly a distributed information system.

Because of the high cost of capturing and preparing geo-spatial data and the high strategic value attached to geo-spatial data, organizations holding this type of data need to exercise access control over its use in terms of who are the authorized users, who can prepare, change or edit the data. Users are categorized according to their rights over the usage of data. Under this classification there are users who can only view and analyze the data without modification and users who can only view the data without analysis and modification. Summarized in the access matrix (Table 1.2) is the user category with their access rights.

Table 1.2 User access matrix

User type/ Access right	Prepare/ create	Edit /modify	Read / visualize	Read/ analyze
Doer/creator of spatial data	Yes	Yes	Yes	Yes
Analyzer/spatial data analyst	No	Yes	Yes	Yes
Viewer/user	No	No	Yes	Yes

Earlier on, the client–server information architecture (Fig.1.4 (b)) was adequate to cater spatial data services to the above types of users. The geo-spatial data produced and validated by a responsible agency are stored in a central server which could be accessed by various types of users in a computing network.

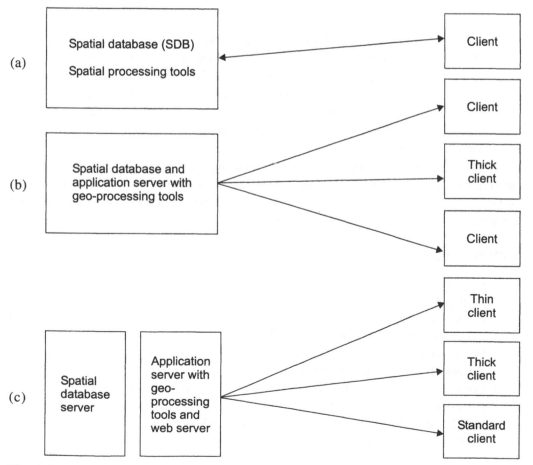

Fig. 1.4 Deployment architecture of GIS as (a) desktop; (b) client–server; (c) enterprise information system

But the client–server information architecture had the following serious lacunas, i.e. often malicious clients could subject the spatial data to unauthorized manipulation. The response time to the user becomes slower when large number of users accessed the same spatial database. Hence the response to user's request was affected. This is known as a load balancing problem in distributed information architecture. The client–server architecture is seriously susceptible to unauthorized use of data and services.

Hence the 3-tier information architecture (Fig.1.4 (c)) has emerged, where the actual data is kept segregated from the user with an application and web server managing the request–response cycle of authenticated users. It authenticates the user before allowing the user access to the system. It also takes care of the load balancing problem by highly optimizing the throughput and response time.

GIS is being used as a sub-system of many operational systems involving different hardware and software with varying architecture. In order to incorporate the rapid developments in research and application in geo-spatial information systems, leading GIS developers have developed a suit of GIS solutions to adapt them to the varying needs of hardware, software and information architecture. Desktop GIS, client–server GIS, web GIS, palm-top GIS, real-time GIS etc. are part of such GIS suits. Besides the above products, a library of GIS objects is being developed to customize and adapt GIS to various target systems. Mobile GIS software—which is being used in mobile hand-held devices such as GPS, mobile survey stations and on-board navigating devices—need to work in real-time operating environment with an inbuilt fail-safe mechanism. The customization and adaptation of GIS to different information architecture will be discussed elaborately in Chapter 11.

The reference model for open distributed processing (RM–ODP) is a systematic way of analyzing different aspects of a system with its relevant perspectives. A RM–ODP analysis of GIS aspects with respect to various perspectives (from the point of view of the planner, designer etc) is tabulated in Appendix A (Table A.1).

References

[1.] Cowen, D. J. GIS versus CAD versus DBMS: What are the differences? *Photogrammetric Engineering and Remote Sensing* 54: 1551–4.1988.

[2.] Goodchild, M. F. Geographical information science. *International Journal of Geographical Information Systems* 6:31–45.1992

[3.] Heywood, I., S. Cornelius and S. Carver. *An Introduction to Geographical Information Systems.* Harow: Longman. 1998.

[4.] Longley, Paul A., Michael F. Goodchild, David J. Maguire and David W. Rhind (eds.). *Geographical Information Systems.* vol 1, vol 2. 2nd ed. John Wiley & Sons. 1999.

[5.] Longley, Paul A., Michael F. Goodchild, David J. Maguire and David W. Rhind (eds.). *Geographic Information Systems & Science.* John Wiley & Sons. 2001

Questions

1. What is GIS? Define GIS in terms of an information system?
2. How is GIS different from a conventional information system?
3. What is spatial data? What is spatio-temporal data? Explain with examples.
4. Explain with diagram the flow of GIS processes.
5. Draw the work flow of a typical GIS system and explain the steps of the flow?
6. What is information architecture? How has GIS emerged through different information architectures?
7. In the context of GIS, what does the acronyms IPO, MVC and DIKD stand for?

2

Input Domain of GIS

Analysis of input domain of GIS is essential to understand the organization and architecture of GIS systems. The set of raw data processed by GIS comprises its input domain. The prime categories of data are spatial, temporal and spatio-temporal in nature. This chapter elaborates on the different forms of spatio-temporal data. The input data processed by GIS as a set, its various forms, sources, agencies and their classification is also discussed. The process flow associated in the pre-processing of the spatial input domain before being analyzed by GIS is described. The physical characteristics of spatial data are discussed to clearly bring out its nature and content.

2.1 Introduction

'What is the input domain of an information system?'—The answer to this question will help to understand what inputs a software or information system can take in for processing. Generally inputs are in the form of data, which are alpha-numeric representation of magnitude of different physical quantities. Data by themselves sometimes cannot convey information unless processed. Processed data conveys the meaning associated with the data, which is called information. In general, software which intake data, process them and imparts information to the data are called 'information systems'. GIS is an information system which essentially processes spatial data to derive spatial information.

2.2 What is Spatial Input Domain?

The potential of an information system in general and GIS in particular can be studied by understanding the input domain it can process. Therefore, the versatility of GIS is directly proportional to the cardinality of the input domain it can process. Therefore, it is pertinent to study the input domain of GIS, i.e. the various aspects of input data such as the content of the data, organization or format of the data, quality, sources and agencies and the way they are modeled for various usages. The input domain of an information system can be formally defined as *'the set of input data and events that it can process to give meaningful information'* (Longley et al. vol. 2, 1999, pp. 665–666).

There is no empirical formula which associates the cardinality of the input domain of a software to its strength and versatility; nevertheless in this chapter, an attempt has been made to portray the strength of GIS through its input domain. It is also important to understand the issues associated with spatial data viz. sources and agencies from where the data is originated, considerations of modeling the digital data for different usage, the quality etc.

Satellite technology has brought in a sweeping change in the way space imaging is done. In tandem with this progress, geo-spatial data capturing has witnessed a phenomenal growth in the frequency at which the image of a particular portion of the earth can be taken with varying resolution. In other words, the frequency (temporal resolution) of capturing spatial data has increased; and so has the spatial resolution i.e. the data obtained can capture in greater detail, the features of the earth surface. To cope with this advancement in data, geo-spatial technology is trying to keep pace by providing powerful spatial processing capabilities, which can handle large volume spatial data for extracting meaningful information. Innovative products such as Google Earth are examples of such systems which has taken Internet users by storm.

A growing input domain means growing areas of applications i.e. a larger user domain. Choosing the appropriate geo-spatial data from a spatial database for a specific application then becomes an issue. The issues that need to be resolved are: what data formats to choose?; what should be the coordinate system of the data?; what geo-referencing system or transformation is to be applied?; what is the geodetic datum to be used? The answer to all such queries can be resolved by creating a meta-data of the geo-spatial database. Of late, preparation of a standard geo-spatial meta-database of the spatial data available has become a national concern. This has been discussed lucidly at the end of this chapter. To bring out various aspects of input spatial domain, it has been enlisted in a tabular form, in a hierarchical block diagram and has been compared and differentiated content- and format-wise.

2.3 Tabular Representation of Input Domain

One way to understand the input domain is to enlist different types of spatial data in a tabular manner (Table 2.1) with the associated source of their origin and the format of data (Longley et al. vol. 2, 1999, pp. 667–675).

Table 2.1 Tabular representation of input domain

Sl. No.	Input data type	Source	Topology/format
1	Raster scanned data	Scanner, unmanned aerial vehicle (UAV), oblique photography	Matrix of pixels with the header containing the boundary information GeoTIFF, GIF, PCX, XWD, CIB, NITF, CADRG

(continues...

Table 2.1 continued)

Sl. No.	Input data type	Source	Topology/format
2	Satellite image	Satellite	BIL, BIP, TIFF
3	Vector map	Field survey, digitizer output	DGN, DVD, DXF
4	Attribute data	Field survey, statistical	Textual records binding several attribute observation, census data fields stored in various RDBMS e.g. Oracle, Sybase, PostgreSQL etc.
5	Elevation data	Sensors, GPS, DGPSLIDAR (hyper-spectral data)	Matrix of height values approximating the height of a particular grid of earth surface. DTED-0/1/2, DEM, NMEA, GRD, TIN
6	Marine navigation charts	Marine survey, coast and island survey, hydrographic and maritime survey organization	S57/ S56 electronic navigation charts, coast and island map data
7	Ellipsoid parameters/ geodetic datum/ geo referenced information/ coordinate system information	Geodetic survey, marine survey, government surveying organization	Topology: Semi-major axis, semi-minor axis, flattening/eccentricity, origin of the coordinate center, the orientation of the axis with respect to the axis of the earth, centered with an earth fixed reference frame
8	Projection parameters	Geodetic survey organizations or agencies	As meta-data or supporting data to the main spatial data—often saved as header information of the main file
9	Almanac and metrological data	Almanac tables	Time of sun rise, sun set, moon rise, moon set, weather information including day and night temperature and wind speed etc.

2.4 Hierarchical Representation of Input Domain

The input domain processed by GIS can be classified into three major categories viz. spatial, aspatial or non-spatial and temporal data. Further, these major categories can be classified into sub-categories and sub-subcategories by the way the data are organized and the sources they are obtained from. The categories and sub-categories can then be depicted in a hierarchical tree structure (Fig. 2.1).

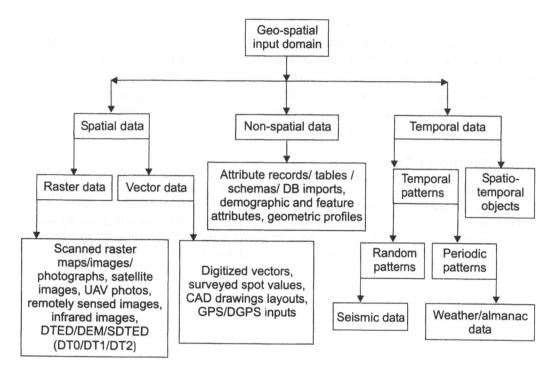

Fig. 2.1 Hierarchical representation of input domain

2.5 Classification of Input Domain (Content-Wise)

To understand how a particular GIS works it is crucial to understand the nature and formats of all the spatial data it can process. Application of GIS is also governed by the spatial data input. New applications are emerging, making the system a tool for collaborative processing, analysis and visualization of spatial data. To keep pace with all these applications, the input domain is also growing. Hence the input domain of a GIS is fairly dynamic in nature. It can be classified in many different ways and context. One way is by the type of association existing among the fields of data or data elements in a record viz.

- Spatial data (Spatial association)
- Temporal data (Association through a time period or epoch of time)
- Attribute data (Association through qualitative property)

2.5.1 Spatial Data

Spatial data types are those, which are associated with the coordinates (x, y) in a 2D-plane or (x, y, z) in a 3D-space. A coordinate imparts a spatial order among geographically occurring objects. Hence all the geographical objects modeled having (x, y, z) coordinates are called spatial data. Physically, spatial data are earth objects, which are geo-coded and geo-referenced i.e. the coordinates of the object are expressed in terms of latitude and longitude in the case of a two-dimensional object and latitude, longitude and height in the case of a three-dimensional object. Further, the earth objects are modeled as point, linear or polygonal objects. One can also define relationships of these entities with other spatial objects in its vicinity such as: touching, contained, intersecting, left of, right of etc. Defining these algebraic relationships among spatial entities is called topology. Topology imparts a spatial ordering among geo-spatial objects. GIS can create a spatial database of earth objects with topology so that spatial analysis can be carried out on these spatial data using conventional geometry.

2.5.2 Temporal Data

Temporal data are spatial data that have a time component attached to it. Therefore, in addition to the coordinate information, a temporal data will have a time component i.e. (x, y, z, t) or (latitude, longitude, height, date and time). The temporal association among spatial objects is quite common. The time component signifies the date and time of occurrence or capturing of the data. These data are known as spatio-temporal data. Spatio-temporal data are important for analyzing change detection, event tracking or temporal behaviour of the geographical phenomena. Temporal modeling and analysis of the geo-spatial data can give an insight into the continuous and discrete spatio-temporal behaviour of the object under consideration (Longley et al. vol. 1, 1999, pp. 91–103). Prominent examples of spatio-temporal data processed by GIS are almanac data, seismic data, weather data etc.

2.5.3 Attribute Data

Attribute data otherwise called non-spatial data are attributes of earth objects which subjectively and objectively describe their nature. They are normal records of any conventional database with the key fields identifying an earth object. There is a one-to-many association of the feature with the data fields. Attribute data are obtained by statistical organizations through various data collection techniques. Attribute data answers the query: What is the spatial object? Popular attribute data processed by GIS are population or census data of a city, mineral deposit data of a state and logistic data of an organization.

2.6 Classification of Input Domain (Organization-wise)

Generally, different GIS systems are strong in capturing, organizing and analyzing a particular type of spatial data. Therefore, they can be classified by the spatial data format they handle. If a GIS is more capable of handling raster data then it is called raster GIS, if it is more oriented to process vector data then it is called a vector GIS. Ideally, all GIS systems should be able to handle all types of GIS data simultaneously—an objective they are moving towards.

Another way of classifying input domain is by the way they are organized, i.e. the format of spatial data. Types of geo-spatial data that are mostly being debated upon by the user community are as follows:
- Raster vs. vector
- Tiled vs. seamless
- Object vs. layer vs. field.

2.6.1 Raster vs. Vector

Whether data should be stored and processed in raster or vector form has historically been a source of much debate. The debate is perhaps dying down slightly because nowadays most GIS systems, at least from a user's perspective, have some provision for handling both. Usually GIS systems compute vector features by converting it to the raster domain and back in different scenarios. It is still true, though, that not all operations of GIS are supported in both types of data.

Raster data of images of earth surface is obtained as an array of pixels (picture elements). Raster data has many advantages over vector data. To name a few, raster data is easily available and easy to display. The operations are typically algebraic in nature involving matrix operations. The simplicity of concepts and algorithms applied to raster data makes it quite analytical. The hardware technology for capture, storage and rendering of raster data has progressed steadily. Remotely sensed data, displayed output, and many digital elevation maps (DEMs) are in raster form. This format is good for data that need to be sampled evenly. Generally, raster data is huge, do not scale well, and introduces aliasing effects when zoomed to high levels.

Vector data is a set of (x, y) or (x, y, z) coordinates describing the features of a terrain. Vector data is obtained by digitizing the prominent features of a raster image. A set of discrete coordinates describes the features. A single coordinate best describes point features such as hut, well etc. A chain of coordinates describe a linear feature such as road, rail etc. A closed chain of coordinates describes an area feature such as agricultural land, playground etc. Vector data is the natural choice for linear features such as roads and boundaries, and mandatory for accuracy in instances of land survey. The speed of vector computation is often criticized; GIS developers and algorithm designers should take this as an opportunity to provide significant improvements. Care should be taken, however, that these improvements will be realizable for 'typical data' and will provide some feature that raster processing does not provide. A complete comparison of various aspects of vector and raster data is given in Chapter 3, Table 3.2.

2.6.2 Tiled vs. Seamless

In vector as well as raster databases, the quantity of data generated becomes overwhelming. Practical considerations dictate that the data be partitioned and considered in smaller chunks. Systems which convert paper maps to GIS, and computer cartography systems are accustomed to dealing with map sheets, quadrants, and tiles. Other users, however, demand a seamless structure—maybe even one that allows a wide range of levels of detail. For various reasons, the earth crust is divided into a uniform grid of tiles. This makes representation easier; handling of data is accurate and simpler. But for the purpose of analysis, sometimes the zone of analysis lies in adjacent tiles forcing the user to represent the data in a continuous and seamless fashion. Hence representation of geo-spatial data as tiles and joining them seamlessly to create a mosaic of the data is important. Spatial partitioning of the data domain into: uniform grid, quad trees, R-trees, binary space partitions etc. and storing them in different data structure for manipulation varies for different GIS.

2.6.3 Object/Layer vs. Field

In the last few years some members of the GIS community have argued that the raster/vector debate obscures more essential issues about how people think about and manipulate space. Consider a situation where you want to describe the agricultural activity in a certain region. There are several spatially referenced variables that may be of interest: the owners, the crops grown, the soil type, elevation, rainfall, etc.

One might speak of farms in the region as geometric objects that have the owner as an attribute. Fields could be objects with the type of crop grown in them as an attribute, and could inherit an owner from the farms that contain them. This is an example where an object-oriented approach to the data is a natural choice. There is clear separation of objects, hence their representation can be encapsulated, and they can form a hierarchy in which representations and attributes can be inherited.

When one adds soil type, however, one may find that a field can contain several soil types or that a soil type may encompass several fields. Sub-dividing fields into objects that can be given a single soil type attribute may or may not be a good idea depending on the importance of the soil type variable. It seems more natural to start a new layer. This in turn creates problems where the same object may have different representations in different layers. A stream, for example, may be represented as a change of ownership and a change in soil type.

It may be more natural to think of soil type, elevation, and rainfall as functions defined over the region—considering the cropland as a 'vector field'. But defining data structures to support continuous data models is difficult because data is usually obtained by discrete sampling. The amount of data needed can be unwieldy and expensive to collect. The traditional soil map quantizes the domain into subjective categories (clay, clay with gravel, sandy, very sandy, etc.) and creates a raster or partitions the region into cells that are assigned one of these soil types to obtain an object-oriented, vector-based map. An implementation of a vector field idea might represent the proportion of each possible component (clay, gravel, sand, etc.) as elements of a vector. One then needs algorithms to quantize this vector space to present understandable output.

Arguments in this debate usually conclude by recommending raster or vector implementation so that the object/layer/field debate, too, becomes limited by current technology.

A map or a globe is a representation of the earth's surface or earth modeled to truly represent the earth features. A model is a representation of the true object keeping all the physical parameters intact. Since maps and globes are prepared in such a manner that the physical parameters are maintained, modeling is an important activity in their preparation with the parameters being different for the preparation of different maps according to their usage and requirement. Some modeling parameters are specific to the patch of earth surface they represent. Therefore a map prepared for navigation may not be valid for land use computation or wild life conservation.

2.7 Sources of Spatial Data

The source from where the spatial data originates plays a key role in its information content. Spatial data obtained from different sources contains different types of information (Longley et al. vol. 2, 1999. pg 655–666). Even data obtained for the same region but from different sources gives a different perspective regarding the region. Therefore, it is often necessary to analyze and study spatial data obtained from different sources together. This needs the data set to be registered and fused to extract the exact information spatially. GIS is capable of integrating spatial and non-spatial data obtained from various sources and agencies and present a consolidated picture of the area under consideration. This is known as spatial data fusion. Figure 2.2 depicts spatial data of different categories obtained from different sources.

Besides the data described in Fig. 2.2, GIS needs the meta-data regarding the categories of data and the geodetic datum they use. The kind of meta-data varies for different types and categories of cartographic data. A suitable candidate list of frequently used meta-data is enlisted in Appendix C, Table C.1.

Besides the traditional geo-spatial data sources, there are many online sources of spatial data available for the purpose of research and development and academic research.

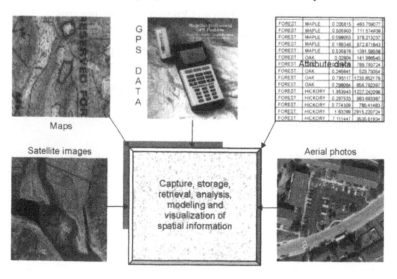

Fig. 2.2 Sources of spatial data

2.8 Process Flow of Spatial Data Capturing

Collection of spatial data is tedious and involves complex systems (Cheng and Lee 2001). The block diagram in Fig. 2.3 shows a generic process of capturing spatial data. The data capturing system may be periodic (e.g. remote sensing satellites) or aperiodic (e.g. unmanned aerial vehicle, UAV), manual or automatic, the process for collection of spatial data follows the sequence of operations as depicted in the block diagram 2.3 below.

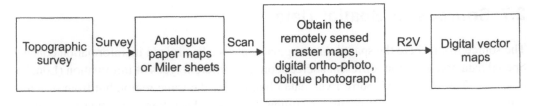

Fig. 2.3 Process flow of spatial data capturing

The sequence of process involved in R2V (raster to vector) conversion can further be depicted through the block diagram given below. One can observe that image processing is used in every step of the process. This is an integral component of any GIS package. There are a number of commercial software available in the market today, which can perform R2V efficiently.

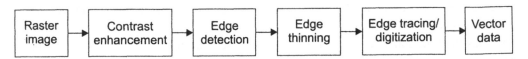

Spatial data belonging to the sea, islands and coastlines is known as maritime spatial data. Maritime spatial data is quite important. It has high economic value and is shared among nations to help in safe navigation. The data is of high volume and often amphibious in nature. Unlike topographic data, maritime data is stored in open format such as S-57 and available in the form of an electronic navigation chart (ENC). The process of capturing maritime data is quite scientific and many sophisticated gadgets are used. Figure 2.4 shows a block diagram depicting capturing of maritime spatial data

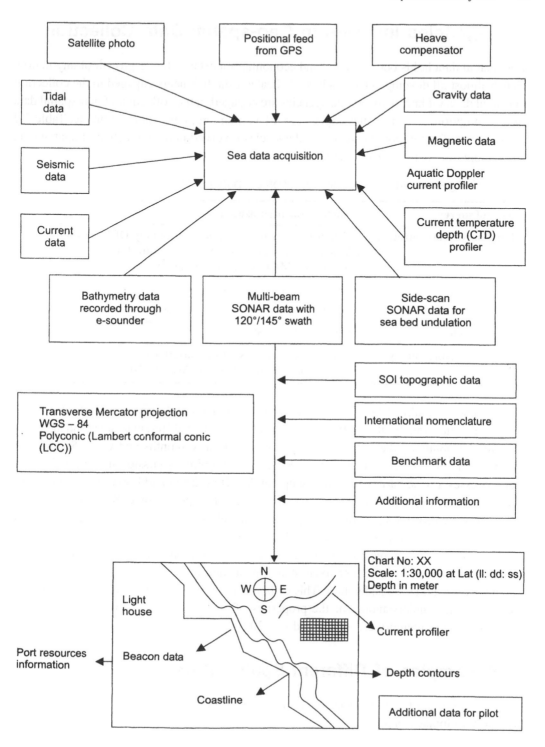

Fig. 2.4 Process flow for capturing maritime spatial data

2.9 Agencies Involved in Geo-spatial Data Collection

Geo-spatial data holds both strategic and economic importance to the growth of any nation. Hence almost all developed nations have dedicated establishments engaged in the collection of spatial data. Of late many private agencies are engaged in the collection of geo-spatial data for consumption by civil bodies engaged in urban planning, tourism, aviation etc. Table 2.2 enlists some of the important sources and agencies involved in the collection of spatial data and building of spatio-temporal database.

Table 2.2 Agencies involved in the collection of spatial data

Type of survey	Name of the organization
Land survey organizations	National survey organizations e.g. Survey Of India (SOI) Military survey organizations e.g. Centre for Automated Military Survey (CAMS)
Hydrographic survey maritime survey	National Hydrographic Survey Organization National Maritime Survey Organization
Topographic survey	Survey of India (SOI)
Remote sensing survey	National remote sensing organizations e.g. National Remote Sensing Agency (NRSA)

The input domain of GIS is characterized by fundamental attributes such as format, scale, and dimension. Further, different data are obtained from different sources and agencies. The time of obtaining these data can vary in terms of days, months and years. Spatial data can be in 2D or 3D, vector or raster etc. From the foregoing discussion it is clear that GIS can processes a wide variety of spatio-temporal data (Fig. 2.5) in addition to the non-spatial data in the form of attributes of the spatial features. The power of GIS lies in extracting information by processing the diverse input domain and making it available for collaborative decision-making. Therefore, choosing a right combination of data (input domain) is important for the success of any project involving geo-spatial information. The correct choice of input data in turn is guided by the functional requirements and the application of the information system project under consideration. This calls for a correlated understanding of individual spatial data i.e. understanding of the properties of spatial data in terms of scale, ground coverage, size in memory etc. (Chen and Lee 2001)

2.10 Samples of Different Spatial Data

The set of Fig. 2.5 below are specimen images of different types of geo-spatial data that a typical GIS intakes as input.

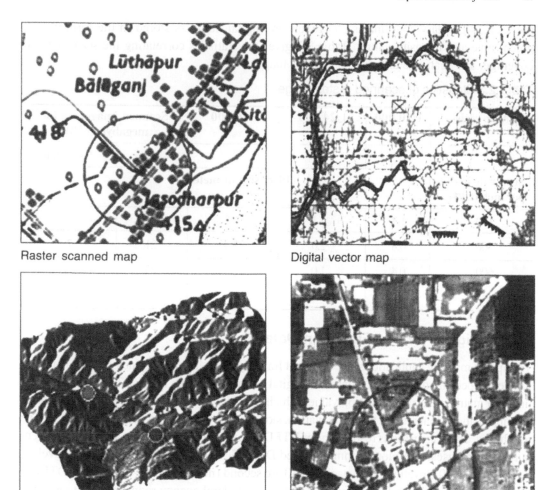

Raster scanned map   Digital vector map

Sun-shaded relief map   Satellite image

Fig. 2.5 Specimen of spatial inputs

2.11 Statistics of Spatial Data

Tables 2.3–2.5 display the statistics of raster, DEM and vector data in terms of the ground coverage, size, scale etc.

2.11.1 Raster Maps

A raster map is generally a digital image manifested in the form of a regular grid of pixels. The dimension of the grid is generally of the size of width × height necessarily preceded by a header which contains the meta-data regarding the image. In case of a scanned map, the contents of the meta-data are the boundary coordinates of the map, the datum and projection information. Besides the above meta-data, there are other data pertaining to the map such as the scale, secondary coordinates etc. Pixels are digital numbers stored in the form of red, green, blue and intensity values. Table 2.3 gives the statistics correlating the scale, ground coverage, and scan rate and size of a raster map

Table 2.3 Statistics of scanned raster image

Map scale	1:10,000	1:24,000	1:40,000	Scanned image size in megabyte (MB)	
				Colour	Grayscale
Scan rate in dots per inch (DPI)	Pixel size in meter				
150	1.7	4.1	6.8	5	2
300	0.8	2.0	3.4	21	7
600	.0.4	1.0	1.7	83	28
1200	0.2	0.5	0.8	334	111

2.11.2 Digital Terrain Elevation Data

To support military applications, the National Imagery and Mapping Agency (NIMA) of USA has developed standard digital datasets—Digital Terrain Elevation Data (DTED®). DTED is a uniform matrix of terrain elevation values which provides basic quantitative data for systems and applications that require terrain elevation, slope, and information regarding terrain surface roughness. There are various resolutions of DTED data available in different levels for different purposes and uses. The most commonly used DTED levels are level 0, level 1 and level 2.

DTED level 0 elevation post-spacing is 30 arc second (nominally one kilometer). DTED 0 was derived from NIMA DTED level 1 to support a federal agency requirement. It was then determined that DTED®0 could be made available (within copyright restrictions) to the public at no charge through the Internet. Support from select international mapping organizations was instrumental in the generation of the level 0 dataset. DTED level 0 may be of value to scientific, technical, and other communities for applications that require terrain elevation, slope, and surface roughness information. It is a gross representation of the undulation of earth's surface for general modeling and assessment activities. Such reduced resolution data is not intended and should not be used for automated flight guidance or other precision activity involving the safety of the public.

DTED level 1 is the basic, medium resolution elevation data source for all military activities and systems that require landform, slope, elevation, and gross terrain roughness in a digital format. It is a uniform matrix of terrain elevation values with post-spacing every 3 arc seconds (approximately 100 meters). The information content is approximately equivalent to the contour information represented on a 250,000 scale map.

DTED level 2 (30 m) or DTED2 is a high resolution elevation data source. It is a uniform gridded matrix of terrain elevation values with post-spacing of one arc second (approximately

30 meters). The information content is equivalent to the contour information represented on a 1: 50,000 scale map.

NIMA has started to develop a strategy to understand and meet the increasing custom demand for higher resolution digital elevation data (DTED). Currently, several private contractors state that they have the capability to collect and generate high resolution elevation data. NIMA is designing a specification criterion to evaluate and test this data when it is made available. The current DTED format does not handle higher resolution data very well, since the finest resolution that can be implemented is a 0.1 arc second. Table 2.4 is a summary of known and proposed post-spacing rows and columns, and tile size. The ground distance is nominal and approximated based on 0–50 degrees N/S latitude.

Table 2.4 Statistics of DTED data

DTED level	Spacing between sampling	Approx. ground coverage in meters	Row × column	Tile size	Scale of map
0	30 arc sec	~1000	1200 × 1200	1 × 1 degree	
1	3.0 arc sec	~100	1200 × 1200	1 × 1 degree	1:250,000
2	1.0 arc sec	~30	3600 × 3600	1 × 1 degree	1:50,000
3	0.3333 sec	~10	900 × 900	5 × 5 degree	

2.11.3 Digital Vector Data

The correlation of various statistics of a vector map is given in Table 2.5. Digital vector data are digitized maps obtained after tracing the prominent features of the scanned maps. The features are digitized coordinate values stored in the form of points, lines and polygons. One can observe that the size of vector data of the same ground coverage is smaller than its raster counterpart because they contain only the skeleton information of the corresponding raster maps.

Table 2.5 Statistics of digital vector data

Map scale	1:50,000	1:250,000	1:500,000
Ground coverage	23 km × 28 km	112 km × 115 km	300 km × 225 km
Ground coverage in spherical coordinates	15' × 15'	1° × 1°	2° × 3°
1 km on ground = x cm on paper map	2 cm	0.4 cm	0.2 cm
Approx. size in memory (in MB)	4	6	10

References

[1.] Bell, S. B. M, B. M. Diaz, F. C Holroyd and M. J Jackson Spatially referenced methods of processing raster and vector data. *Image and Vision Computing* 1:211–20. 1983.

[2.] Chen, Yong-qi and Yuk-cheung Lee (eds.). *Geographical Data Acquisition*. New York: Springer Wien. 2001

[3.] Davis, Raymond E., and Francis S. Foote *Surveying Theory and Practice*. 4th edition. McGraw-Hill.1953.

[4.] Jones, Christopher B. *Geographical Information Systems and Computer Cartography*. Longman. 1997.

[5.] Longley, Paul A., Michael F. Goodchild, David J. Maguire and David W. Rhind (eds.). *Geographical Information Systems (Principles & Technical Issues)* vol. 1, vol. 2. 2nd edition. John Wiley & Sons. 1999.

[6.] Longley, Paul A., Michael F. Goodchild, David J. Maguire and David W. Rhind. *Geographic Information Systems & Science*. John Wiley & Sons. 2001.

[7.] Peuquet, D. J and L. Qian. An integrated database design for temporal GIS, Seventh International Symposium on Spatial Data Handling, Delft, Holland. International *Geographical Union* 2:1–11. 1996.

Questions

1. What is the input domain of an information system?
2. What is geo-spatial input domain?
3. Classify the input domain of GIS?
4. What is R2V? Describe the R2V process with a block diagram.
5. Describe the process flow of preparing a marine chart?
6. What are the constituents of a marine chart?
7. What kind of data is required to carry out the expansion plan of your university campus and why?
8. Give the generic contents of the input domain of GIS with examples.
9. Describe the process flow of spatial data capturing, explaining each sub-process involved.

3

Spatial Data Modeling

Having understood what the input domain of GIS is and its various aspects, it is essential to represent them in an empirical form for further processing in computing machines. In GIS terminology, this is often termed as pre-processing of input data or spatial data modeling. Spatial data modeling is important for visualization and analysis of geo-spatial data (Clarke 1990). Modeling imparts syntax to geo-spatial data in various ways thus preparing geo-spatial information so that a common user can infer the spatial information along with the context. In traditional cartography, numerous ways of spatial data modeling were developed to represent a patch of earth on a paper map. Modern cartography has also developed many modeling techniques to print, visualize and analyze geo-spatial data in a digital environment. Of late, temporal modeling of geo-spatial data to extract terrain change is gaining momentum and has many potential applications. This chapter starts with an answer to the question: What is geo-spatial data modeling? It then goes on to explain the various types of spatial data models used by GIS, the key factors influencing the spatial data model and the steps required to design a spatial database. The temporal modeling of spatial data is discussed with its various aspects. Finally the meta-data required for extracting and understanding spatial data from the spatial data repository is discussed.

3.1 What is Spatial Data Modeling?

A model is a mathematical or visual representation of a physical object and its associated events. The characteristics of the physical object are represented using data and their behaviour emulated through suitable mathematical equations. Geo-spatial objects are modeled using the spatio-temporal data pertaining to them and the various operations that can be performed on them to extract different aspects of geo-spatial information. In order to perfectly model geo-spatial objects the following key aspects of the terrain surface are preserved viz.

1. The physical representation of the object in terms of point, line or polygon
2. The *inter se* distance, direction and orientation of the object with respect to its surrounding objects.

3. The overall position, length, perimeter, surface area, volume and shape of the object.
4. The graphical and symbolic representation of the terrain object, which will give a visual representation to the objects in terms of shape, colour, style, pattern etc.
5. The scale, horizontal and vertical datum, projection parameters and the coordinate system associated with the geo-spatial data.
6. How the data is stored or organized in the memory and how it is accessed using logical data structure. Suitable spatial file formats which is the outcome of the spatial data organization and data structure that can hold the data while computation is also preserved.
7. The temporal behaviour of the spatial features i.e. the behaviour of the object with respect to time e.g. change in shape, attribute pattern etc.

Modeling spatial data involves storing it into different data structure. Data structure imparts syntax to the spatial data. The operations that can be performed to manipulate and extract information from the data structures are well-defined in terms of algorithms in computing science. In other words, storing spatial data into a data structure means imparting syntax to the spatial data (Frank 1992). A GIS aims at processing, managing and analyzing spatio-temporal data. The capability of any information system in general and GIS in particular largely rely on the design of its data model. A data model presents the conceptual core of an information system. It defines the data object types, relationship among data structures, operations that can be performed and the rules to maintain database integrity (Date 1995). A rigorous data model must anticipate the spatio-temporal queries and analytical methods to be performed in the GIS. Information regarding temporal constructs must be represented by data objects defined in the data model of a GIS. Because of the temporal analytical capability, many GIS systems are called temporal GIS.

3.2 Geo-Spatial Data Modeling

Earth surface, which is generally referred to as 'terrain surface', contains numerous features both natural and man-made. The frequency and occurrences of these features differ in different regions (e.g. coastal, mountainous, desert etc.) and time. In a way, the occurrence of a particular feature on terrain depends on both space and time and hence is a spatial and temporal phenomenon. Modeling terrain features involves giving a uniform definition so as to describe the complete attributes of every feature i.e. its spatial occurrences, its spatial and temporal relationship with its surrounding features, its appearance etc. To represent the feature symbolically so as to uniquely visualize and identify it in a physical container viz. a paper map or digital map or an elevation model etc also forms a part of the spatial modeling known as cartography.

Modeling of earth features in terms of spatial, temporal and attribute data and quantifying them as digital information is an important prerequisite for processing GIS data. The organization and storage of these digital information guides the kind of data structure needed to hold them and the type of algorithm to be applied for retrieval, processing and display by the GIS. Saving the digital information in various data formats so as to retrieve, modify and update for different purposes is quite intriguing and poses lots of challenge in designing GIS systems (Frank 1992).

For the purpose of making maps, the earth surface is divided into various zones with each zone having uniform spheroid or ellipsoid parameters. Each zone is further sub-divided into sub-zones and a mosaic of squares known as grid or graticule. Each square is surveyed (locating of features relative to each other) using location information with respect to an origin of the graticule. The surveyed terrain feature is modeled keeping in view its usage (e.g. preparation of paper map, printing of paper map, creation of thematic map, topological analysis of earth features, viewing of terrain model etc). Depending on different usage of the geo-spatial data, modeling differs and hence the way the data is being organized, stored and further analyzed.

The most important geo-spatial data models prevailing are vector, raster or DEM (digital elevation model) (Carter 1988). Table 3.1 lists some aspects that have to be considered while modeling terrain features for different usages. As can be seen in the table, all modeling considerations are not applicable to all sets of geo-spatial data.

Vector data model envisages the object as a set of coordinates (x, y, z). These data are stored along with the associated datum parameters and are assigned with different colour, style, line type and width to represent them uniquely in a digital display or paper container. The raster data is a digital image displayed in the form of a matrix of the pixels representing the patch of earth. Each pixel is encoded as red, green, blue and intensity modeling the optical characteristic of the object. The digital terrain elevation data models the earth surface as a grid of square patches. Each patch is assigned the average height of the patch from mean sea level. Hence DEM is essentially a matrix of height values with a header containing the meta-data regarding the terrain it represents.

Table 3.1 Spatial data model and their modeling considerations

Sl. No.	Spatial data model	Modeling considerations
1	Vector model	(x, y, z) coordinates: for retrieval of location information
		Colour, style, line type/pattern, symbology and symbol size: To impart a unique visualization to the feature
		Major category, minor category, sub-sub category: For thematic composition of maps, categorization of features into major and minor categories; to create thematic and operational maps
		To archive, query, organize, save, update, and retrieve from digital data
		For measurements of earth features e.g. area, distance, perimeter and volume
		For analysis e.g. to compute shortest path, optimum path, surface area, cross country mobility condition etc
		Preparation of paper map for printing

(continues...

Table 3.1 continued)

Sl. No.	Spatial data model	Modeling considerations
2	Raster model	For printing and visualization of maps in digital display
3	DEM (digital elevation model)	(x, y, z) coordinate for location in 3D-space, for a 3D-perspective view of terrain surface, fly through and walk through visualization
		To compute optical and radio line of sight
		To compute elevation value for visualization of terrain surface

Each of these models has its strength and weakness. A comparison of the prime spatial data models with different aspects brings out the strengths and weakness of the models for different applications. It gives the builder of spatial information resource a general idea as to what types of modeling criteria to choose for building different applications.

Beside these primary terrain data models, there are many secondary or derived models. These are specific to applications which make use of them. Some prominent secondary spatial data models are contour models, grid models and TIN models. A contour model is used for terrain surface visualization. A grid model, which is a derived form of a raster model, is used for modeling digital terrain elevation and visualization of surface. The triangular irregular network (TIN) model is used for realization of spatial query such as point in an area or space, range query etc. Each of these models can be derived from one another through highly optimized and established algorithms.

3.3 Raster vs Vector Data Model

One of the most important debates among geo-spatial community is the raster vs. vector model argument. A typical GIS user is always concerned about what advantage, disadvantage he or she gets by adapting a particular spatial data model functionally. It is because a particular set of functions, which is applicable over vector model, may not be applicable to raster model and vice versa. Table 3.2 gives a comparison between the vector and raster model of geo-spatial data with different perspectives, which will set to rest some of these concerns.

Table 3.2 Raster vs. vector data model

Property/ perspective	Raster	Vector
Spatial feature representation	Grid of pixels or cells Point = A single cell Line = A series of connected cells Area = A region of contiguous cells	Series of coordinates Node = Single coordinate Line = Two coordinates Polygon = Series of line segments with the first and last coordinate the same

(continues...

Table 3.2 continued)

Property/ perspective	Raster	Vector
Coordinate representation	Location of each cell is identified by the grid's origin plus the relative horizontal and vertical distance of the cell from the origin	The basic unit of representing geo-spatial information in the vector model is the coordinate. The coordinates are a pair or triplet of numbers expressing the distance from a set of orthogonal axis e.g. 2D/ 3D and rectangular or polar coordinate system. The data are represented in the form of (x, y, z) or (easting, northing, altitude) or (range, bearing, phi) or (latitude, longitude, altitude)
Explicit attribute	The attribute of each cell are expressed using red (R), green (G), blue (B) and intensity (I) values which are the spectral measures corresponding to the patch of earth	Attributes of each object are attached as a record stored in RDBMS. Hence each object has a link to a record in the RDBMS
Spatial resolution	The resolution is expressed as the minimum linear distance/ dimension of the smallest feature, which can be identified in the image. High resolution means smaller dimension of the pixel. Low resolution means larger cell dimension and less information content	The resolution of vector data is expressed in the form of a scale. The scale is a ratio of the distance of two points on the digital map to the actual ground distance. Higher the scale, lower is the precession. Lower the scale, higher is the precession. 2 cm on a digital map = 2 km on the ground implies that the scale of the map is 1:50,000
Data structure used	Matrix or a 2-dimension array of values storing the R, G, B and I per pixels organized as bytes per pixels, pixels per rows and rows per segments and number of inter-connected segments as image files	The data structure used to store a vector representation is an array, quad tree and RB tree.
Operations and functions that can be applied	A convoluting matrix can be used to the entire image matrix, which apply the average value to the value of the center cell, or the median value, or apply a dominance rule etc. Sometimes the value of each cell is checked for a threshold to apply binary rules such as zero or one thus making a binary image.	Topologies i.e. the relative positioning of spatial objects are easy to represent in a vector model. One can prepare a TIN (triangular irregular network of nodes) and thus the following operations can be applied in vector data: Line–line intersection Point inside triangle/polygon

(continues...

Table 3.2 continued)

Property/ perspective	Raster	Vector
	The outcomes are edge prominent images; noise removed images; binary images; grey-valued images	Compute the convex hull of the object To find the nearest location of a particular object etc.

Figure 3.1 shows the visual results of a geographical space using both vector and raster data model.

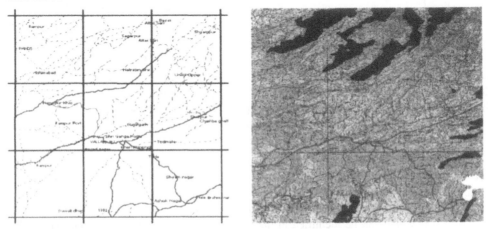

Fig. 3.1 Vector and raster representation of the same geographical area

3.4 Process Flow of Spatial Data Modeling

The process flow usually adopted in any GIS for modeling spatial features is given below in the form of discrete steps as well as a block diagram (Fig.3.2).

STEP 1   Earth feature (real) → Photo/scan (matrix of digital values expressed in R, G, B and I)

STEP 2   Digitization (R2V) → Vector/geometrical/graphical representation (colour, line type, style and width)

STEP 3   Projection + coordinate system + datum → Point, line, polygon representation

STEP 4   Thematic categorization (all similar sets of earth features as category of features) i.e. thematic layer or thematic composition of similar features

STEP 5   Establish the relationship of the features using various geometrical associations e.g. adjacent to, by the left of, contained inside etc. i.e. applying a topological relationship among the digitized features e.g. TIN, RB tree, quad tree, oct tree etc.

STEP 6   Representation of the vector data in a higher level of abstraction i.e. class representation of the geo-spatial data which means spatial data and the functions that can be applied on the spatial data. Instances of spatial class are manifested as a spatial object with different spatio-temporal events acting on the object for effective visualization or analysis.

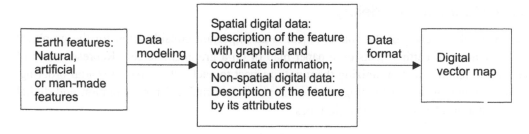

Fig. 3.2 Process flow of geo-spatial data modeling

Figure 3.2 shows the various steps and processes involved in the preparation of a digital map data from an earth feature. Thus, the representation of the earth feature is finally saved in efficient data structures and data formats. The data structures and formats are modeled keeping in mind the need for ease of storage, retrieval, updation and visualization of the geo-spatial data. Other factors include portability across various operating systems, hardware platforms, application software; and inter-operability across various visual and analytical software. Keeping all the above factors in mind, earth features are modeled as a spatio-temporal object having certain essential attributes (spatial, non-spatial and temporal) describing every aspect of the feature (Worboys 1994). Thus, as explained in previous chapters, the generic representation of a feature can be described in three categories i.e.

1. Spatial attributes (E, N and A) answering the query: where the feature is located on the earth?
2. Non-spatial attribute retaining the property of the earth feature (a, b, c,..f..) answers the query: what the object is about?
3. Temporal attributes (dd-mm-yy, hh-mm-ss), answering the query: when the object has occurred?

Therefore, satisfying all the above aspects of the earth object, a generic template can be designed to describe it (Fig. 3.3).

Fig. 3.3 Template of spatial feature

In Fig. 3.3, data is modeled in the form of a class with the attributes as its member data. Different terrain objects can thus be modeled with different values for the member data.

3.5 Different Types of Spatial Modeling

There are various ways of modeling spatial data. Prominent among them are (i) geometric modeling; (ii) spatial modeling and (iii) thematic modeling

3.5.1 Geometric Modeling

Geometrical modeling of the spatial object is an important consideration for the graphical representation of earth objects on a paper map or on a graphical screen. Representation of geometric features and organizing them with respect to each other gives a spatial ordering. This is known as spatial topology. Table 3.3 gives the graphical representation and digital representation of various geometric types.

Table 3.3 Cartographic modeling of spatial data

Geometric type	Graphic representation	Digital representation and topology
Point		(x, y) in 2D, (x, y, z) in 3D
Line		(x_i, y_i), $i = 1, 2, \ldots$ in 2D (x_i, y_i, z_i), $i = 1, 2, \ldots$ in 3D
Area/polygon		(x_i, y_i), $i = 1, 2, \ldots n$, such that $(x_1, y_1) = (x_n, y_n)$
Surface		Rectangular grid of height values. TIN (triangular irregular network) $f(x, y) = z$, contour representing surface of equal height.
Volume		Set of bounded surfaces $f(x_i, y_i) = z_i$

3.5.2 Thematic Modeling

A group of terrain features clustered together according to a common property or theme they share is known as thematic categorization or thematic modeling of terrain features. In thematic modeling, terrain objects are stored as records having some important property. A collection of such records having a common attribute value is known as thematic layer. Many such thematic layers constitute a map stored as a file or table in spatial database. It helps the surveyor, cartographer and GIS software designer to collect, collate and analyze spatial data thematically for the following purposes:

1. To store as digital data
2. To print or plot a paper map with clear and distinct colour and legends
3. To store and retrieve or display the digital map by a GIS software
4. To query a set or category of features
5. To create a thematic map for a particular purpose
6. To represent ground features as digital features using special symbols
7. To view a map in a different scale and analyze the data

The properties described in the template can uniquely represent a spatial feature and help GIS software to model the data to view the feature in various scales; to plot/print; to query; to organize; save/update/retrieve or display as digital data. The template can be implemented as a structure in most programming languages or as member data of a class in object-oriented languages by representing it as a class with each feature as an object (instance of the class) as given in Fig. 3.4.

Every feature on the earth's surface/terrain can be categorized into some theme as given in the second column of Table 3.4. The categories vary from region to region. Further each category is divided into sub-categories as in column three of Table 3.4. These features are further classified according to the geometric shape they best represent i.e. the topology of the feature, for the purpose of modeling (e.g. point, line, polygon or area and text) and representation in a digital map.

Table 3.4 The spatial feature category and sub-category with topology

Sl No.	Category/theme	Sub-category represented in map as	Topology
1	Buildings	Buildings (residential)	P
		Buildings (religious)	P
		Antiquities	P
		Buildings (others)	P
2	Hydrographic	Rivers	A
		Canals	L
		Other water features	A
		Hydro associated features	A
		Coastal features	A/P
3	Communication	Roads	L
		Railways	L
		Embankments and cuttings (roads and railways)	P
		Aerodromes	P
4	Land covers	Land cover	A
		Vegetation	A
		Land use features	A
5	Utilities	Transmission lines	L
		Oil pipelines	L
		Water pipelines	L
6	Boundaries	Administrative divisions	A
		Administrative boundaries	A
		Limits	A
7	Hypsography	Contours	A
		Mountain features	A
		Mud volcanoes	A

(continues...

Table 3.4 continued)

Sl No.	Category/theme	Sub-category represented in map as	Topology
		Sand features	A
		Heights, BNs and control points	P
		High mountain features	A
8	Vital installations	Civil vital installations	P
		Military vital installations	P
9	Map frame and text	Marginal and border items	L
		Names	T
		Grid (metric)	T
		Grid (FPS)	T

```
Class feature {
    Category                // Major type [XX YY]
    Feature name            // Sub-category name
    Abbreviated name        // to be referred to by database as key name
    Feature code            // [00–99] Category (XX) + Sub-feature (01-YY)=XX YY
    Feature type            // Topology of the feature e.g. Point, Line, Polygon or Text
    Level code              //[0–63] Layers into which the feature belongs to
    Line code               //[0–9] 0–8 for map features, 9 for map frame and text
    Weight code             //[0–6] maps to [0.75 mm–0.25mm–0.8mm] to represent
                            font
    Colour code             //[0–27] Which colour the feature will be represented by
    Digitizing command      // The feature is digitized as e.g. Cell, Symbol, Line, String
    Remarks                 // Digitizing command if any
    DVD code                // A combined code to attach a non-spatial/attribute data
                            // table to the feature [11–63][XX YY] Unique code for the /
                            / feature. Digital vector data code.
                            // Some of the member functions can be
        Display_feature(feature name)
        Retrieve_feature(feature name)
        Print_feature (feature name, level code, colour code)
        Query_attribute(feature name, DVD code)
} // End of class definition for feature
```

Fig. 3.4 Class representation of the feature template

Any terrain feature can be manifested as an instance of the class shown in Fig. 3.4, i.e. as an object that can be manipulated individually or combined with other features.

Each feature when associated with a spatial coordinate i.e. (latitude, longitude and altitude) or (easting, northing and altitude) becomes a spatial feature. This spatial feature can further be associated with the time of survey or occurrence—if it is associated with the time factor, it is called a spatio-temporal feature. Of late most GIS software gives a facility to link vector data with its associated records containing the attribute values through a common field linking the two.

Map making is a major design consideration of the above data model. On a paper map, different features can only be optimally represented by different thickness of lines, points, polygons or text. Hence different weights are assigned to different sets of topographical features so as to obtain the required clarity at the time of plotting, printing or display. Cartographers are advised to follow the criteria given in Table 3.5 to obtain a good paper map.

Table 3.5 Cartographic modeling of spatial features

Sl. No	Line weight (Wt)	Line thickness (in mm)
1	Wt = 0	0.075
2	Wt = 1	0.100 0.125
3	Wt = 2	0.150 0.175
4	Wt = 3	0.200 0.225
5	Wt = 4	0.250 0.275
6	Wt = 5	0.300 0.325
7	Wt = 6	0.350 0.375

In Table 3.5 each line weight code represents two line thickness except the first one (line weight = 0) which is taken as default while digitization. The second needs to be assigned explicitly by the cartographer. The convention for line thickness should be followed for all linear and text features appearing under categories 1–6 of Table 3.4. However for features appearing under category 9, i.e., map frames and text annotations, the convention mentioned in Table 3.6 should be followed.

Table 3.6 Line thickness for map frame and text

Sl. No	Line weight	Point size (inch/per—I/R)	Line thickness (in mm)
1	Wt = 0	6I/R	0.175
2	Wt = 1	8I/R	0.275
3	Wt = 2	9I/R	0.325
4	Wt = 3	12I/R	0.450
5	Wt = 4	14I/R	0.525
6	Wt = 5	18I/R	0.650
7	Wt = 6	24I/R	0.800

3.5.3 Spatial Modeling

Spatial characteristics of geographical features can be broadly classified into attributes which describe where the feature is by means of spatial coordinates, reference positions, spatial units and spatial relationships with other features (Mitasova et al. 1995). In a sense, spatial modeling imparts a relative positioning of the features with its neighbouring features. This is sometimes known as building topology.

Spatial location information is captured using the position of the feature with some coordinate systems, whereas spatial relationship is described using *inter se* distance, orientation from north or relative angle to an established feature. Spatial information of a feature predicts its position with respect to other referenced features in the database.

Modeling spatial information must take into consideration the object, as it is; the presence of the object with respect to other objects; and finally the occurrence of the object in a larger topographical area. The relative positioning of the object is modeled using network models such as terrain irregular network (TIN) or Voronoi tessellation. The continuity property of the feature can be modeled using a field model. A geographical information system (GIS) must take into consideration the above three modeling considerations for representing geographic information as given in Table 3.7.

Table 3.7 Models that can be used to represent geographical information

Sl. No	Type of model / conceptual view	Type of information	Examples
1	Object based model	Discrete features with all attributes	Building, survey tree,
2	Network model	Interconnection among individual features	Interconnecting road, rivers, drainage system
3	Field model	Continuous phenomena	Property of the feature as a function of the continuous location e.g. plain area, dense forest, desert etc.

3.6 Temporal Modeling

The development of temporal modeling of GIS has a parallel to the progress in temporal data modeling in computer science (CS). In CS the temporal date model was first incorporated in relational models and then in object-oriented models. In GIS the first temporal model was implemented with a time stamping of each thematic layer. This was then extended to time stamping of events or processes. Data semantics is the key idea in modeling and the fundamental idea is to raise the level of abstraction in the transition from layers/tables to events/objects. Temporal modeling of data gives the much-needed analytical capability to detect the temporal changes of earth features, change of patterns and behaviour of many natural phenomena which are crucial for scientists and researchers. Hence almost all emerging GIS are incorporating many temporal analysis methods and are called temporal GIS.

Representation of time is a crucial factor in temporal GIS (Armstrong 1988). Time can be considered as the fourth dimension to the spatial data. Usually spatial data is considered in 2D (planimetric data), 3D (complete representation of spatial data in space). The fourth dimension can be spatio-temporal representation such as (pi, ai, gi, ti), where (pi, ai, gi, ti) corresponds to the position, attribute, geometry and time of occurrence or temporal description of the feature. From the time point of view, objects can be categorized into (i) static and (ii) dynamic objects. Static information in the real world are cartographic maps, roads, railway tracks, public utilities etc. which may not change in a short period of time—all these objects will change during a long period of time. Dynamic information refers to that category of geo-spatial objects that change in a short period of time. Hence the length of the period is the key criteria for modeling temporal behaviour.

According to the working domain, one can categorize dynamic aspects of spatial data under spatial change pattern and temporal duration pattern. Table 3.8 describes various types of temporal modeling and some relevant examples to make various aspects of the temporal model clear.

Table 3.8 Types of spatio-temporal data modeling

Temporal aspect	Change type	Example
Geometric change	Geometrical changes of features over time	Urban expansion pattern, deforestation, coast line change patterns etc.
	Positional change of features over time	Movement of vehicle, glacier movement, sea wave pattern etc.
	Attribute change in features over time	Traffic rate during different times of day
	Any combination of the above three	Rainfall measurement in different seasons
Duration of occurrence	Real-time spatial data (Real-time refers to the capability of the GIS to visualize and analyze spatial attributes as soon as they are input into the GIS)	Rate of traffic pattern visualization, tracking of vehicle movement, moving map display in cockpit etc.
	Near real-time spatial data (Momentary updating, visualization and analysis of spatial data that need intermediate processing steps before it could be used by the GIS)	Traffic patterns, crime patterns, flood patterns etc.
	Time stamped spatial data (Time stamping of spatial data refers to the time attached to the data during the event of capturing the data)	Time of occurrence of the real world event; duration of occurrence when first brought to GIS environment or received etc; survey date, date and time of photograph etc.

Various types of spatio-temporal data modeling techniques can be categorized according to the types of syntax they impart to the data and the GIS operations that can be performed on them. Figure 3.5 depicts various modeling techniques adopted in geo-spatial data modeling. Almost all advanced GIS systems adopt most of these modeling techniques.

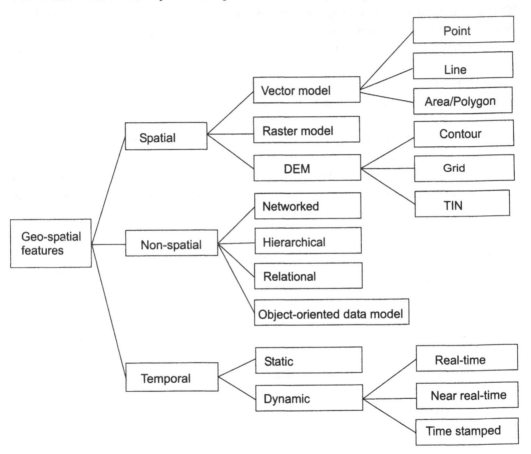

Fig. 3.5 Geo-spatial data modeling techniques

3.7 Spatial Database Design

The next logical step of spatial data modeling is to store all the features in the form of a digital file in mass media, or relational database or as spatial database (Peuquet and Quan 1996). Arranging correlated spatial data in a pre-defined way gives rise to spatial data formats. Data formats play a predominant role in deciding the data structure and ultimately the design of the GIS software under consideration. Various spatial data formats are available. Each has its own design consideration, proprietary consideration, advantages and disadvantages. The more the number of data formats a GIS system can support, the more versatile and popular it is.

The popularity and usage of some spatial data formats are due to the ease of inter-operability among various spatial processing applications and users. Some formats are specifically devised to share data among a group of organizations and industry and hence are patronized by a consortium of academic institutions, industry and government agencies.

An important activity of any GIS project is to create a geo-spatial database. The geo-spatial input domain is pre-processed and need to be stored in some traditional relational database for exploitations. Hence one of the pre-conditions of any project is to process the spatial data so as to create a spatial database. This database design plays a major role in the success of the GIS project. A sound spatial database design is a foundation to efficient storage of query, analysis and visualization of spatial data. The following distinct steps are involved in the creation of a spatial database:

STEP 1 Delineating the study area spatially (identification of area of interest).

STEP 2 Identification and collection of spatial data from various agencies such as survey organizations, digital mapping centers, remote sensing agencies, census handbook etc.

STEP 3 Deciding upon the resolution and accuracy need of the spatial data i.e. topo-sheets of 1:250,000, 1:50,000 scale maps.

STEP 4 Creation of maps using cartographic techniques.

STEP 5 Creation of digital maps and DEMs, which involves the following:

- Digitization of scanned maps, or satellite images for extraction of earth features and assigning colour, style, line and pattern to each digitizing element.
- Editing of the coverage to remove digitization errors such as dangles (isolated nodes), overshoots or undershoots and incorrect labels for each of the polygons.
- Categorization and sub- and sub-sub-categorization of the digitized features i.e. thematic composition of similar features
- Creating topology by performing and establishing spatial relations among digitized features such as contained in, left neighbour, right neighbour, inside etc.
- Applying spatial projection and transformation to the spatial data

There are a number of ways an earth feature can be referenced and searched from a spatial database. It is therefore important for the designer of the spatial database to optimize and create data infrastructure in such a way that different segments of the user community can search, retrieve and analyze the data easily. This calls for spatial database normalization which involves spatial indexing of the data, removing the redundant key and transitive dependency of the fields. One of the important outcomes of spatial database design is the consolidation of the key attributes of the spatial data. The collection of such key attributes regarding spatial data is known as meta-data.

3.8 Spatial Meta-data

Meta-data is data about data, i.e. Meta-data gives information regarding the meaning of actual data (Maguire et al. 1991, Chapter 10). Generally spatial data are complex and huge. Therefore to understand this kind of data, its context i.e. the syntax, semantics and some key information regarding the data need to be understood. Meta-data answer these queries regarding actual data:

- Where does the data belong?
- How is the data organized?
- When was the data collected?
- How is the data to be used?

A user of GIS needs to know the precise answers to the above queries before making a decision regarding choice of appropriate spatial data required for the geo-spatial application under consideration.

From the discussions in Chapter 2, it can also be concluded that the input data domain is diverse in many respects and is continuously evolving. The spatial data itself has high strategic and economic value while the user domain of the spatial data is spread across the globe and is multilingual.

Therefore, keeping in the view the diverse nature of the spatial data, a standard set of meta-data need to be designed (Maguire 1991, Chapters 10, 30) so as to make meaningful reference to the diverse spatial data together through some key characteristic of the spatial data domain, required by the user. Figure 3.6 further elucidates the above discussions regarding meta-data.

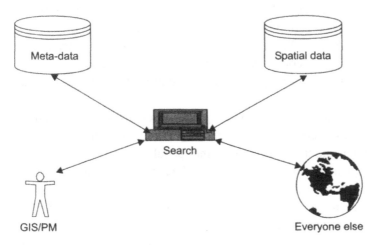

Fig. 3.6 Searching through spatial meta-data

3.8.1 Definition of Spatial Meta-data

At this juncture it is apt to enlist the definition and function of geospatial meta-data as spelt out by the ISO.

> Spatial meta-data encapsulates knowledge 'about the identification, the extent, the quality, the spatial and temporal schema, spatial reference, and distribution of digital geographic data'.

ISO-19115 (2003)—Geographic Information—Meta-data

After defining geo-spatial meta-data, a natural question arises: 'What are the functions of geo-spatial meta-data?'

'The objective of the geo-spatial meta-data, is to provide a structure for describing digital geographic data so that users will be able to locate, access, evaluate, purchase, and utilize spatial data, determine whether the data in a holding will be of use to the application under consideration'.

ISO-19115 (2003)—Geographic Information—Meta-data

3.8.2 Key Functions of Geo-Spatial Meta-data

For a lucid understanding of the definitions given in the preceding paragraphs, it is pertinent to understand and discuss further the key functions of geo-spatial meta-data and some important or most frequently used meta-data. Enlisted below is an illustrative list of spatial meta-data functions:

1. Provide data producers with appropriate information to characterize their geographic data properly.
2. Facilitate the organization and management of meta-data for geographic data.
3. Enable users to apply geographic data in the most efficient way by knowing its basic characteristics.
4. Facilitate data discovery, retrieval and reuse. Users will be better able to locate, access, evaluate, purchase and utilize geographic data.
5. Enable users to determine whether geographic data in a holding will be of any use to them.

3.8.3 Key Elements of Spatial Meta-Data

Some of the core meta-data elements adopted by numerous geo-spatial data producing organisations are as follows (Maguire et al. 1991, Chapter 30)

- Dataset title (Name of the data set e.g. MAP of INDIA)
- Dataset reference date (Release date)
- Geographic extent (4 boundary coordinates or coordinates of the polygon enclosing the data)
- Dataset language (English/French/Japanese etc.)
- Dataset character set (English/Devanagari/Chinese/...)
- Dataset topic category (Cadastral/Communication/...)
- Spatial resolution (1:50,000, 1:250,000, ...)
- Abstract
- Distribution format (TIFF, GIF, JPEG, ...)
- Spatial representation (GRID, TIN, ...)
- Reference system (WGS84, NAD, ...)
- Lineage statement
- Online resource
- Date and time of collection
- Source and agency responsible (Name of organization having ownership of the data)

The meta-data table in Appendix C gives a matrix of meta-data pertaining to different kinds of spatial data obtained from different sources and agencies.

Reference

[1.] Armstrong, M. P. Temporality in spatial database. Proceedings: GIS/LIS'88, 2:880–889. 1988.

[2.] Carter, J. R. Digital representation of topographic surfaces. *Photogrammetric Engineering and Remote Sensing* 54:1577–80.1988.

[3.] Clarke, K. C. *Analytical and Computer Cartography*. Englewood Cliffs: Prentice-Hall. 1990.

[4.] Date, C. J. *An Introduction to Database Systems*, 6th ed. Edition-Wesley. 1995.

[5.] Frank, A. U. Spatial concept, geometric data models, and geometric data structure. *Computers and Geosciences* 18:409–17.1992.

[6.] Houlding, S. *Three-dimensional Geosciences Modeling*. Berlin: Springer.1994.

[7.] Maguire, D. J., M. F. Goodchild and D. W. Rhind (eds). *Geographical Information Systems: Principles and Applications*. Harlow, UK: Longman. 1991.
Chapter10: Generalizing spatial data and dealing with multiple representations.
Chapter 30: Generalization of spatial database
Chapter 49: Metadata and data catalogue.
(Text available online at www.wiley.com/gis and www. wiley.co.uk/gis)

[8.] Mitasova, H., L. Mitas, W. M. Brown, D. P.Gerdes, I. Kosinovsky and T. Baker. Modeling spatially and temporally distributed phenomena: New methods and tools for GRASS GIS. International Journal of Geographical *Information Systems* 9:433–46.1995.

[9.] Peuquet, D. J. and L. Qian. An integrated database design for temporal GIS. Seventh International Symposium on Spatial Data Handling, Delft, Holland. *International Geographical Union* 2:1–11.1996.

[10.] Worboys, M. F. A unified model for spatial and temporal information. *The Computer Journal* 37:26–33.1994.

Questions

1. What is a model? What aspects of spatial data are considered for modeling geo-spatial data?
2. What are the important geo-spatial data models?
3. Enlist the differences between raster and vector data models?
4. Describe the processes involved in spatial data modeling with a block diagram and explain its sub-processes.
5. List the different types of spatial data models.
6. What is geometric modeling? Enlist the geometric model types with an example for each.

7.   What is thematic modeling?
8.   Give class representations of an earth object in object-oriented paradigms.
9.   Describe the steps involved in designing a spatial database?
10. What is meta-data? What is spatial meta-data?
11. What are the functions of geo-spatial meta-data?
12. Describe some key elements of geo-spatial meta-data.

4

Processing Tools of GIS

The importance and popularity of a GIS system depends on how capable the GIS system is in providing processed information according to the user need. The ease at which processed information is provided is crucial to its popular usage. The processing capabilities of a GIS are often manifested through sophisticated geo-processing tools. These geo-processing tools consist of various geometric calculations, geodesic computations, astronomical formulae and spatial data transformations. Mathematical transformations such as coordinate transformations, map projections, geometric transformations, geodesic calculations and datum transformation are highly specific to geo-spatial data processing. These tools are manifested through complicated mathematical formulae, which can be solved in many different ways. The governing parameters of these equations are mapped to physical parameters specifying the geo-spatial objects or data under consideration. Finally these formulae are translated through various algorithms, which can be set as a computer programme using different programming languages.

This chapter explains the need for geo-processing components and their categorization according to the functions and applications they cater to. Further it explains the mathematical equations that go into building commonly used GIS tools. Spherical trigonometry, which is used in computing applications related to astronomical objects, is discussed in the end of the chapter.

4.1 What are the Processing Tools in GIS?

A geo-processing tool is an algorithmic implementation of a computer programme manifested as an application programmed interface (API), dynamic linking library (DLL), component object model, distributed component object model (COM/DCOM) or some other computing component available in the GIS. In other words, a geo-processing tool is an implementation of an algorithm applied on a data structure holding spatial data. It is capable of processing and extracting meaningful geo-spatial information. The process of transforming a computing formula to a geo-processing tool is depicted in the block diagram as given in Fig. 4.1.

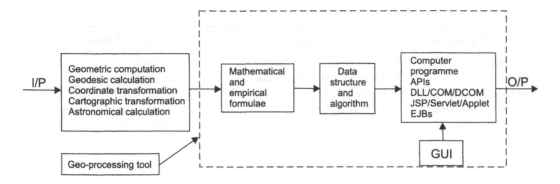

Fig. 4.1 Process flow of geo-processing tools

All these complex processing are abstracted and offered to the user through a very powerful, yet simple to use graphical user interface (GUI). In fact GUI of a GIS has helped in increasing its popularity among the user domain. GUI is an important criterion in the selection of any GIS. The processing tool kit of GIS systems differs from one system to another.

One way of benchmarking the efficiency of any GIS is by assessing how optimized and efficient the geo-processing tools it offers are. Generally optimization is achieved in terms of the processing time and the amount of memory the programme consumes. Different GIS vary in how efficiently these geo-processing algorithms are implemented in terms of processing time, speed and data handling capability etc.

GIS tools transform geo-spatial data from one form to another, giving better information qualitatively, quantitatively and visually. Without processing the geo-spatial data, this information is not evident in the actual raw data. Thus by processing the data, the user can comprehend geo-spatial information or draw better inferences from the processed spatial data.

4.2 Categorization of GIS Tools

GIS tools can be categorized depending on the type of geo-spatial data they can operate upon. Table 4.1 enlists a few common geo-processing tools that are available in a GIS. The prime categorizations of these geo-processing tools are as per data they process and the output they produce. They can be categorized as 2D analysis, 3D analysis and visual analysis.

Table 4.1 Different types of analysis in GIS

2D analysis	3D analysis	Visual analysis
Locate (ϕ, λ) on map	Find height from MSL at given (x, y) or (ϕ, λ)	Generate a 3D perspective view of the terrain

(continues...)

Table 4.1 continued)

2D analysis	3D analysis	Visual analysis
Compute: Distance Shortest distance Geodesic distance Crow fly distance between two points (ϕ_1, λ_1) and (ϕ_2, λ_2)	Find (easting, northing and altitude—ENA from MSL) of a given point (ϕ, λ) To compute ortho-morphic height of a place.	Generate: Sun-shaded relief Colour coded relief 3D perspective view of the terrain surface
Compute: Slope Aspect Height At a point on earth surface.	Compute: Surface area of a piece of earth Volume of a piece of land from mean sea level (MSL)	Generate: Wire mesh Fly through Walk through Of the terrain surface
Compute: Perimeter Surface area of a polygonal region	Compute slope and aspect of a point Find the minimum and maximum altitude of a polygonal zone Find the azimuth and elevation of a celestial body	Compose and display thematic map Generate land usage map Generate political map Generate cadastral map
Find and locate a geographical feature on the map	Compute line of sight between two points on the earth's crust	Fly through simulation Walk through simulation

4.2.1 2D Analysis

Analysis in two dimensions involves measuring and quantification of the spatial and non-spatial attributes of an earth feature or set of features. The set is defined by applying aggregate of topology or statistical method on the spatial data. The quantification of the attribute under consideration is done in the GIS by taking the two-dimensional spatial geometry of the digital data. Table 4.2 shows a sample list of 2D analysis and its corresponding result types.

Table 4.2 2D Analysis and results in GIS

	2D analysis function	Purpose/result
1	Finding the (easting, northing) of a cursor point on the map in different precision i.e. E(2)N(2), E(3)N(3) digits etc.	Easting and northing in meters
2	Finding the latitude and longitude in degree, minute and seconds in DD: MM:SS	Finding the latitude and longitude in DMS of the point.
3	Finding the length between two points on a map marked by a cursor	Length between two points on a map in meters

(continues...

Table 4.1 continued)

	2D analysis function	**Purpose/result**
4	Finding the cumulative length between successive points on the map.	Cumulative distance of any two points along with the intermediate points on the map.
5	To compute the perimeter of a polygonal zone.	Perimeter of the area feature represented by a closed chain of vectors.
6	Finding the surface area of the polygon defined on the map.	Surface area in square meters

4.2.2 3D Analysis

Analysis in three dimensions involves quantifying the attribute/attributes of features on earth represented by the digital data identified by the volume or geometry of R^3 or in space. Volumetric measurement of quantity involves the x, y and z coordinate of the map. A representative list of 3D analysis components found in GIS is enlisted in Table 4.3.

Table 4.3 3D analysis and results in GIS

SN	**3D analysis function**	**Purpose/result**
1	Computing the ENA of the cursor position	Computing the latitude, longitude and ENA of any position on earth
2	Finding the minimum and maximum altitude of a polygonal zone	Computing the altitude from MSL at any point on earth
3	Finding the line of sight (LOS) between two points	To locate the ideal point for establishing communicating points e.g. transponders and antennas.
4	Finding the LOS fan (explained in Section 4.3.5) of a point	To find the ideal location for establishing communicating elements e.g. transmitter and receivers.
5	Finding the volume of the polygon defined by a user	To compute the mass of earth or water content of the defined polygon.
6	Slope and aspect of each point	To find the drainage pattern of the land
7	Fly through and walk through	To simulate terrain visualization; to carry out reconnaissance of a hostile terrain or unknown terrain

4.2.3 Visual Analysis

Cognitive analysis or visual analysis is an inbuilt capability of a human being. By this process, human beings see, observe and think about the feature simultaneously and extract required information. After observation, a human being comprehends approximate information regarding the object. GIS gives a formal platform to visually analyze the terrain information. The visual analysis is carried out through various 2D and 3D visualization tools offered by GIS. In GIS, maps are prepared according to specific themes e.g. cadastral maps, population maps, land use maps, political maps and elevation maps. Each of these thematic maps conveys different information in the form of colour, texture or annotation, which are given along with the map as legends. These are known as digital visual maps and can be prepared by applying hybrid queries raised by the user. These queries are formed based on spatial, non-spatial/attribute and temporal behaviour of the earth feature or earth objects. Examples of such queries resulting in viewing map for analysis are given in Table 4.4.

Table 4.4 Query analysis and results in GIS

SN	Query	Thematic map type	Observations
1	Display all national and international boundaries including boundaries of states and major grids	Political map	Size and area of each state. Size and extent of countries. The biggest and smallest state/country in terms of size.
2	Display the places where ores are available (iron ore in red dots; bauxite in pink, gold in green).	Ore mapping	Availability of iron, bauxite and gold. Which ore is available in maximum quantity in a specific area?
3	The basin of river Ganga and Yamuna in summer, winter and rainy season.	Cadastral mapping	The basin of river Ganga and Yamuna. Area cf the basin under flood zone during rainy season. Area of the basin where cultivation can be done during summer season.

Besides the visual analysis of 2D geo-spatial data, 3D perspective viewing of the terrain surface gives the user immense potential to analyze the pattern of terrain. 3D perspective viewing is carried out on digital elevation data viewed from different user perspectives to see the undulated surface as it is. Sometimes the earth surface is viewed after generating a sun-shaded relief map or moon-shaded relief map to depict the terrain elevation. Shading is done on the surface of the earth taking into consideration the azimuth and elevation of the sun or moon. One can see the relative elevation from the viewer's eye, the slope and aspect of the earth surface from a 3D perspective. This viewing analysis is quite comprehensive and gives the user a natural view of the ground under consideration. An extensive discussion on this topic is given in subsequent chapters.

4.3 Computations Involved in GIS

Calculating the location, height, area, perimeter, volume, slope, aspect, distance between two points, line of sight between two points on earth surface etc are some of the most fundamental and common computations in GIS. Some advanced GIS even compute the azimuth and elevation of celestial objects from the earth's surface and the almanac data pertaining to every location on the earth for a given date and time. Sometimes an application-specific GIS can compute path profile between source and destination, crest clearance of a projectile, elevation profile of a terrain cut and radio line of sight between source and destination. Generally a GIS tool kit should have some of these computing tools in various forms. In this section we will discuss the numerical formulae of a representative set of such tools. It is not exhaustive as the set is ever increasing with more sophisticated spatial computations emerging day by day.

4.3.1 Computation of Height

Computing the height from the MSL (mean sea level) of a given location when its location coordinates (latitude, longitude) are known, has many applications. This is basically the height or elevation of a point on the earth's surface and is defined as the vertical distance from the datum surface to the undulated surface (Fig. 4.2). Almost all global positioning systems (GPS) measures the height of a point. Often the geoid surface is considered as the datum surface. The height of a point above the geoid is known as the orthometric height. Orthometric height can be positive or negative depending on whether the point is located above or below the geoid surface. In case of the height when obtained through the GPS/ DGPS it is computed from the reference ellipsoid (a mathematically-defined surface such as WGS84, that approximates the geoid). Therefore the height obtained from GPS and DGPS are known as ellipsoidal heights. An ellipsoidal height can be positive or negative, depending on whether the point is located above or below the surface of the reference ellipsoid. Ellipsoidal heights are of purely geometrical value rather than having any physical meaning. Modeling of an earth shape to a geoid or an ellipsoid gives two different surfaces. The geoid–ellipsoid separation is known as the undulation of the earth surface or the geoidal height. The relation between the orthometric height, ellipsoidal height and the geoidal height is given below:

Orthometric height (MSL) = Ellipsoidal height +/– Geoidal height

Orthometric heights are in a way synonymous to height from MSL, which is used more frequently. Geoid and ellipsoid height is used in survey, aviation and astronomy.

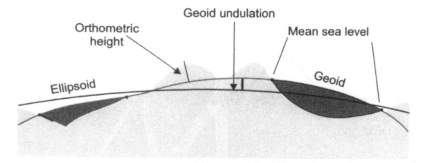

Fig. 4.2 Relation between the height from geoid, ellipsoid and orthometric surface

4.3.2 Computation of Surface Area of a Polygon

Area is a fundamental parameter derived from terrain analysis, which is necessary for numerous decision-making processes. Generally distance and area are the fundamental terrain parameters associated with spatial data. They find many applications while supporting decision-making involving spatial data such as area analysis and cadastral applications. In most GIS, distance and area calculations are based on the vector data model, which makes a planar approximation of spatial data. In reality, the distance or area needs to be computed for an undulating surface. Given below is an approach to compute true area and distance from spatial vector data when the scale and projection are given. The spatial data is integrated with the elevation model to compute slope and aspect of each and every point, thus preserving the undulated property of the surface.

This technique utilizes digital terrain elevation model (DTEM) data in addition to the polygon layer of the vector data to extract the surface area. A DEM is a matrix representation of the continuous variation of the relief of the earth surface over space. In another sense, DEM is a continuous grid containing the elevation at its spatial location in each grid cell. Usually a DEM is interpolated from surveyed digitized contour data, spot heights, and triangle heights. But more accurate DEM data is extracted from stereo aerial photographs or stereo digital remote sensing imagery.

Computing the surface area from the slope gradient involves computing the slope area for each grid point. The surface area of the patch is the result of the aggregate of the surface area due to each grid.

The slope area for each grid can be approximated from the resolution of the grid and the slope value for the grid. In Fig. 4.3, area ABEF is the slope area for the grid ABCD. The slope area for this grid is computed by the expression:

$$\frac{AB \times CD}{\cos(\theta)}$$

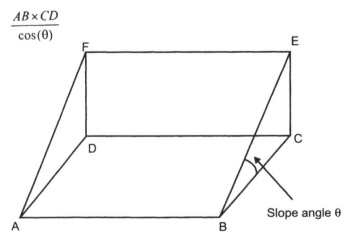

Fig. 4.3 Slope area of a DEM grid

Sometimes the area to be computed is marked on the map as a polygon represented by a closed chain of segments as depicted in Fig. 4.4. The polygon is then decomposed to a set of non-intersecting triangles using well-known polygon triangulation algorithms. Hence the area of the polygon is the sum of the areas of the triangles and is computed using equation (4.1). Suppose the polygon is represented by a series of points (x_i, y_i), $i = 1, ...n$, where n = number of points representing the vertices of the polygon then the area of the polygon is given by

$$A = \frac{1}{2} \left| \sum_{i=1}^{n+1} (x_i y_{i-1}) - (x_{i-1} y_i) \right| \tag{4.1}$$

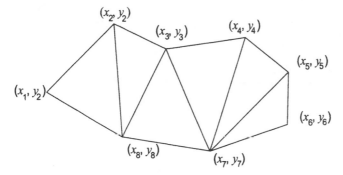

Fig. 4.4 Triangular decomposition of polygonal area

4.3.3 Computation of Volume

Computing the volume of a given surface area and depth marked on a 2D map view or a 3D perspective view of terrain has many applications. Not many GIS available today can compute the volume of a chunk of earth surface from 2D spatial data containing contours. Contours are iso-lines representing equal height on the ground. Generally they are concentric circular patches tagged with a height value. Computing volume from the contour data of a digital map involves the following steps.

STEP 1 The ground (defined by surface area and depth) for which volume has to be computed is cut into pieces along the contour planes. This results in a series of horizontal slabs as depicted in Fig. 4.5.

STEP 2 Each slab is treated as a prismoid with the height equal to the contour interval and the end areas enclosed by the contour lines.

STEP 3 The volume of each prismoid is calculated.
Volume of the prismoid between two contours C_i and C_j

$$= CI \times \left[\frac{A_{ci} + A_{cj}}{2} \right]$$

where CI is the contour interval i.e. the interval between consecutive contours.

STEP 4 The volume of the land mass is the sum of the intermediate prismoids and is obtained by totaling the volume due to these elements.

i.e., Volume between C_i and C_j, is the volume due to the intermediate prismoids plus the volume due to the tip of the contour:

$$V_{ci-cj} = \frac{CI}{2} \times \left[A_{ci} + 2(A_{ci+1} + A_{ci+2} + \dots + A_{cj-2} + A_{ci-1}) + A_{cj} \right]$$

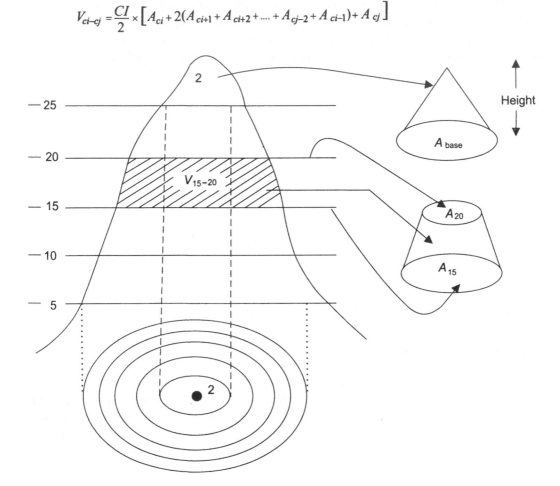

Fig. 4.5 Computation of volume using contour data

4.3.4 Computation of Slope

Computing the slope and aspect at each and every point on the surface of the earth represented by a digital map requires the height of the earth surface at each and every point. Hence to compute the slope, digital elevation data (DEM) of the patch of earth is used. DEM is nothing but a matrix of height values. An algorithm which is one frequently used will be discussed in this section. It computes the slope at each and every point of the matrix by taking into consideration the height value of its neighbouring pixels.

Consider the 3×3 matrix of height values of a DEM corresponding to a map. The height cells are named corresponding to the middle cell as depicted in Fig. 4.6. The computation or window operation described below has to be applied to the complete DEM representing the entire map along the width and height of the map. This is known as neighbourhood window operation and is a well-established scientific principle applied in image processing and signals processing. The neighbourhood operations differ according to the number of neighbour cells taken for computation of slope. Given below is the computation for four and eight neighbour cells respectively. The accuracy and outcome of the result is more when the operating window size increases.

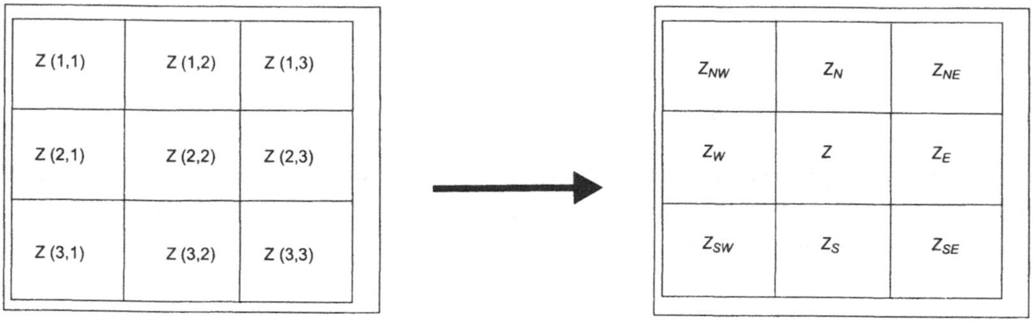

Z (1,1)	Z (1,2)	Z (1,3)
Z (2,1)	Z (2,2)	Z (2,3)
Z (3,1)	Z (3,2)	Z (3,3)

Z_{NW}	Z_N	Z_{NE}
Z_W	Z	Z_E
Z_{SW}	Z_S	Z_{SE}

Fig. 4.6 A 3×3 DEM matrix with corresponding height designates

In Fig. 4.6, Z is the height at the middle cell and Z_E, Z_W, Z_S and Z_N are the height of the East, West, South and North cells respectively.

Similarly Z_{NW}, Z_{SW}, Z_{NE} and Z_{SE} are the height of the Northwest, Southwest, Northeast and Southeast cells respectively. Let the *inter se* distance between two adjacent cells = D

The slopes along the X and Y direction as depicted in Fig. 4.6 can be computed using the set of partial derivatives given in equation (4.2) and (4.3) respectively

(i) Using Four Neighbour Cells:

$$\frac{\Delta Z_X}{\Delta X} = \frac{Z_E - Z_W}{2D} \text{ and } \frac{\Delta Z_Y}{\Delta Y} = \frac{Z_N - Z_S}{2D} \tag{4.2}$$

(ii) Using Eight Cells:

$$\frac{\Delta Z_X}{\Delta X} = \frac{Z_{NE} + Z_E + Z_{SE} - Z_{NW} - Z_W - Z_{SW}}{6D}$$

$$\text{and } \frac{\Delta Z_Y}{\Delta Y} = \frac{Z_{SW} + Z_S + Z_{SE} - Z_{NW} - Z_N - Z_{NE}}{6D} \tag{4.3}$$

The slope at location (x, y) is computed with the partial derivatives obtained from equations (4.2) and (4.3) using equation (4.4).

$$\text{Slope} = \sqrt{\left(\frac{\Delta Z_x}{\Delta x}\right)^2 + \left(\frac{\Delta Z_y}{\Delta y}\right)^2} \tag{4.4}$$

The slope value is more accurate when eight points of the window is used rather than four points.

Aspect is the orientation of the slope in degrees. In an aspect map, each cell is assigned an aspect value, which tells the direction (north, east, south etc.) to which the slope is oriented. Aspect is computed using the partial derivatives given in equation (4.5), where the earth surface is modeled by the function $f(x, y) = h$ (height from MSL).

$$\text{Aspect}(\alpha) = a \tan 2^{-1} \frac{\frac{\delta f}{\delta x}}{\frac{\delta f}{\delta y}} \tag{4.5}$$

The rate of change of slope i.e. the first order derivative of the slope or the second order derivative of DTM gives the curvature of the earth surface. Generally curvature describes the terrain surface in terms of how convex or concave or plane the surface is with respect to its surroundings. Curvature is computed using second order partial derivatives as given by equation (4.6)

$$\text{Curvature} = \sqrt{\left(\frac{\delta^2 f}{\delta x^2}\right)^2 + \left(\frac{\delta^2 f}{\delta Y^2}\right)^2} \tag{4.6}$$

The slope, aspect and curvature of a particular patch of earth surface are used to determine the patterns of flow of water and flow acceleration; to detect terrain change; and to evaluate land for different purposes.

The above formulae applies for DTM data where the height values are represented as a matrix representing the discrete terrain surface with a height of 15 ×15 meter or 30 ×30 meter square modeled as a pixel or cell.

4.3.5   Line of Sight Computation

Computing the line of sight between the observer and the object has lots of application in civil as well as military applications. One of the prime applications of GIS in the communication industry is to identify a suitable location on terrain for locating and installing a communication transponder so as to achieve maximum visibility or reach of receiving elements. This calls for computing whether the source can view the object from its position or not (forward LOS). Reverse LOS computes whether the object under consideration can view the observer.

Computing line of sight (LOS) between two points (observer and object) also has wide applications in operation planning in the armed forces. In a graphical representation, LOS is depicted as a coloured line with green as the visible portion and red as the invisible portion of the line. With no intervening obstacles, the optical line of sight (OLOS) can be expressed as:

$$D = 3.57\sqrt{H} \tag{4.7}$$

where D = distance between the source and the object on terrain surface in kilometers,
$\quad\quad H$ = height of the observer in meters from the MSL

In this equation, it is assumed that there is no intervening (crest) terrain between the source and the observer.

The same can be extended to compute the radio line of sight, which is given by equation (4.8)

$$D = 3.57\sqrt{KH} \tag{4.8}$$

where K is an empirical constant, which adjusts for refraction of radio waves. Generally $K = 4/3$.

The objective of these computations in GIS is to compute the height of the observer or antenna to be installed so as to view a particular target or range of the terrain. This leads to computing the line of sight fan (LOS fan) whereby the observer tries to see the entire terrain surrounding it by rotating its eye 360 degree around its position. Here the visible areas are highlighted in green and invisible areas are highlighted in red.

Inversely the distance between two communicating elements e.g. antennas can be computed using the formula given by equation (4.9)

$$D = 3.57\left(\sqrt{KH_1} + \sqrt{KH_2}\right) \tag{4.9}$$

where $K = 4/3$, H_1 and H_2 are the heights of the two antennas

Hence to compute the optical LOS, the height profile of the intervening terrain between the source and observer is computed for each point with a uniform range. The height profile is compared against the straight line joining the observer and the object point by point. If at any point, the height of terrain is more than the line joining the object and the observer then an LOS does not exist.

4.3.6 Shortest Distance between Two Points

Computing distance between two points on a map has many connotations. The distance can be the aerial shortest distance (crow fly distance); the shortest distance as depicted by a line joining the two points drawn on flat surface; the cumulative distance obtained by taking the segment distance along the digitized vector joining the two points; or the shortest distance out of many paths among the paths joining the two points. One possible computation is the shortest distance between two points P_1 (ϕ_1, λ_1) and P_2 (ϕ_2, λ_2) on the earth surface modeled as a spherical surface (Fig. 4.7).

To find a suitable mathematical formula to compute the shortest distance, slice the spherical earth along the two points and the center of the spherical earth. One can imagine that the slice is a circle with the center same as the center of the earth and P_1 and P_2 lying on its circumference. The shortest distance will be the arc length of the circle passing through these two points. If the radius of the spherical earth is R then the spherical distance D is given by equation (4.10)

$$D = R\cos^{-1}(\sin\varphi_1 \sin\varphi_2 + \cos\varphi_1 \cos\varphi_2 \cos(\lambda_1 - \lambda_2)) \tag{4.10}$$

where D = spherical distance between point P_1 (ϕ_1, λ_1) and P_2 (ϕ_2, λ_2)
R = the radius of spherical earth

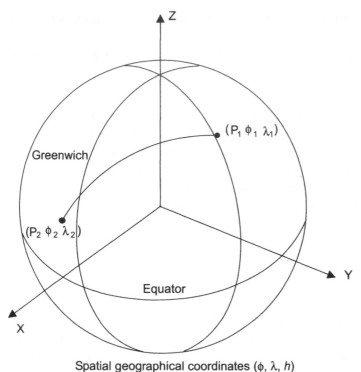

Spatial geographical coordinates (ϕ, λ, h)

Fig. 4.7 Distance between P$_1$ (ϕ_1,λ_1) and P$_2$ (ϕ_2,λ_2) on the earth surface

4.4 Spherical Trigonometry

Imagine a slice of earth surface along any three non-collinear points on the earth surface; it results in a spherical triangle as depicted in Fig. 4.8. The relation between the length of sides and the spherical angles of the spherical triangle leads to computation of many features on the earth surface and hence has many applications in GIS. The solution of a spherical triangle depends upon the principles of spherical trigonometry, of which the cartographer or surveyor should have some knowledge. A derivation of the fundamental equations of spherical trigonometry follows:

In Fig. 4.8, let OX, OY and OZ be the X, Y, Z axes of the rectangular coordinates, and let ABC be a spherical triangle on the surface of a sphere of unit radius of which O is the center, the side c being in the XY plane.

Since the radius of the sphere is unity, each of the distances OA, OB, OC is unity, and the arc angles a, b and c are measures, respectively, of the center angles BOC, COA and AOB respectively.

Let H mark the projection of C on the XY plane and let JH be constructed parallel to OY. Then, since it is parallel to the YZ plane, DCJH is equal to DA of the spherical triangle ABC. The coordinates of C are $x = $ OJ, $y = $ JH, and $z = $ HC. Then, since the radius of the sphere is unity

OJ = x = cos b
JH = y = sin b cos A
HC = z = sin b sin A

Similarly

cos a = cos b cos c + sin b sin c cos A
sin a cos B = cos b sin c – sin b cos c cos A
sin a sin B = sin b sin A

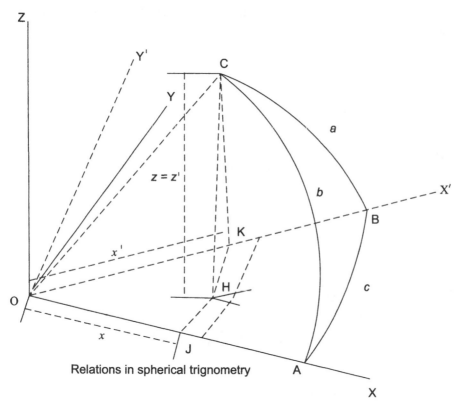

Relations in spherical trignometry

Fig. 4.8 Spherical triangle

4.5 Computation of Azimuth and Elevation of Celestial Body

The visual influence of a celestial body on the earth surface is quite important and has many applications. The relief of the earth surface can be visualized using the shadow created on the surface due to the lighting of the sun or moon. Hence sun-shaded relief or moon-shaded relief can give a simplistic idea of the undulation of the earth surface. These relief views of the earth crust depend upon the lighting effect produced by the sun or the moon, which in turn depends upon the azimuth and elevation of the celestial object at a particular time. Hence computation of azimuth and elevation of a celestial object is important to the modeling and visualization of earth.

In many advanced GIS, computation of almanac data of any given coordinates (latitude, longitude, date and time) finds a place among advanced options. Natural shading of a terrain while fly through or walk through simulation is an important function of many GIS. To simulate the lighting due to the sun or moon, one has to find the position of the sun or moon with its azimuth and elevation with respect to the observer's position, which again calls for the computation of the azimuth and elevation of the celestial object.

Figure 4.9 depicts the upper half of the celestial sphere in which O is the observer and the earth considered to be a point object in the center of the celestial sphere. To compute the azimuth and elevation of any celestial object with respect to an observer standing on the earth crust consider

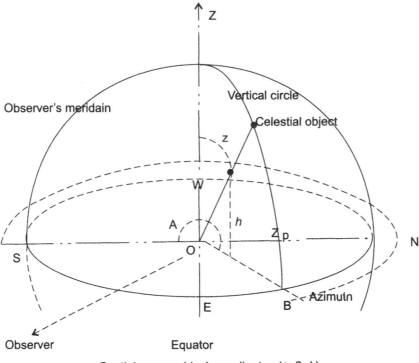

Spatial geographical coordinates (ϕ, ?, h)

Fig. 4.9 Celestial hemisphere depicting observer and celestial object

NESW in Fig. 4.9 is the observer's horizon, and SZN is the meridian plane passing through the location of the observer. Point Z of the celestial sphere is directly above the observer and is called the zenith point. The point C represents a celestial body, and BSZ is part of a great circle called the vertical circle through the body and the zenith. In this horizon system of spherical coordinates, the angular position of a celestial body is defined by its azimuth and altitude.

The azimuth of a celestial body is the angular distance measured along the horizon in a clockwise direction from the meridian to the vertical circle subtended by the celestial body.

Azimuth can be measured from the south or north point of the meridian. For astronomical computations, the azimuth is computed from the south, whereas for navigation and survey, the azimuth is computed from the north.

The altitude of a celestial body is the angular distance measured along a vertical circle, from the horizon to the body. It is expressed in degree of arc. In Fig. 4.9, the altitude is given by angular distance h subtended by angle COB. The complement of the altitude is called the zenith distance or co-latitude. It is the angular distance from the zenith to the celestial body measured along the vertical circle. In the figure, zenith distance is ZOC = $90° - h$. Generally the azimuth and altitude of a celestial object keeps changing continuously because of the relative motion.

At a known instance of time when the celestial body is observed, the altitude and the horizontal angle of the celestial body is measured with respect to the observer's meridian. Then a celestial or astronomical triangle can be formed with the celestial object, the observer and the zenith point as the coordinates (Fig. 4.10).

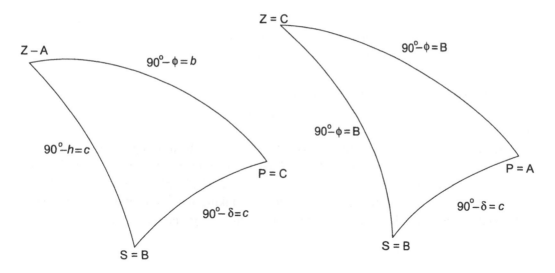

Fig. 4.10 Celestial triangle

Solution of PZS triangle:
In surveying, the astronomical triangle is solved in connection with the azimuth. Observations are made on the sun or on some star that can be readily identified. The altitude of the celestial body is measured; its declination at that instant place of observation is either known or is determined by separate observation. Hence, the three sides of the astronomical triangle are known. The determination of azimuth of the celestial body involves the computation of the angle at Z; and determinations of longitude or time involve the computation of the angle at C as a measure of the hour angle.

In Fig. 4.10, PZS is the astronomical triangle whose sides are $90° - \theta$ (the co-lattitude), $90° - h$ (co-altitude or zenith distance), and $90° - \delta$ (the co-declination or polar distance). If the spherical triangle is rotated in position so that its version A, B, and C coincide, respectively,with Z, S and P of the astronomical triangle, then $a = 90° - \delta$, $b = 90° - \phi$, $c = 90° - h$, and A = Z. Using these values we get equation (4.11).

$$\cos z = \frac{\sin \delta}{\cos h \, \cos\phi} - \tan h \tan\phi \qquad (4.11)$$

which is a general expression for determining azimuth from the north when the three sides of the astronomical triangle are known, Z being considered positive if the star is east of the meridian and negative if the star is to the west of the meridian. It is greater than 90° if the sign of cos Z is positive and lesser if the sign is negative.

When azimuths are measured from the south, the equation (4.11) takes the following form

$$\cos A = \tanh \tan \phi - \frac{\sin \delta}{\cosh \cos \phi} \qquad (4.12)$$

in which the azimuth measured from south is clockwise or counter-clockwise depending on if the celestial body is leaving or approaching the meridian. It is greater than 90° if the sign of cos A is found to be positive and lesser if the sign is negative.

Equation (4.12) may also be expressed in terms of the versed sine (1 minus the cosine), as follows:

$$\text{vers } Z = \text{vers } p - \text{vers } (\phi - h)\sec\phi\sec h \qquad (4.13)$$

where $p = 90° - \delta$ = polar distance.This is a convenient form when tables of versed sines are available.

Alternatively, an equation simplyfing equation (4.13) can be developed for the unknown angle at P by assuming the vertices A, B, and C of the spherical triangle as given in Fig. 4.10 to coincide with the P, S and Z vertices of the astronomical triangle respectively. As shown in Fig. 4.10 , $90° - \delta = a$, $90° - \phi = b$, $90° - h = c$. Making these substitutions in equation (4.11) and letting P = t, the hour angle in either direction from the meridian, one obtains

$$\cos t = \frac{\sin h}{\cos \delta \cos \phi} - \tan \delta \tan \phi \qquad (4.14)$$

which is the general expression for determining the hour angle of any celestial body when the three sides of the astronomical triangle are known. Equation (4.14) can also be expressed in the form:

$$\text{vers} t = \sec\phi\sec\delta \; [\text{vers } z - \text{vers}(\phi - \delta)] \qquad (4.15)$$

where $z = 90° - h$ = zenith distance.

The azimuth of a line can be determined at a single observation of the sun anytime when it is visible, provided the latitude of the place is known.

The sun is observed at a known instance of time, and the altitude of the sun and the horizontal angle from the sun at that given instant is found from a solar ephemeris (Table containing the azimuth and elevation of the sun at a given latitude and longitude). With the declination δ, latitude ϕ, and altitude h known, the PSZ triangle is solved by equation (4.16) given below.

$$\cos z = \frac{\sin \delta}{\cos h \cos \phi} - \tan h \tan \phi \qquad (4.16)$$

where z is the azimuth of the celestial object from north. The azimuth A from the south is given by equation

$$\cos A = \tan h \tan \phi - \frac{\sin \delta}{\cos h \cos \phi}$$

Reference

[1.] Worboys, M. F. GIS: *A Computing Perspective*. London: Taylor & Francis. 1995.

[2.] Coppock, J. T. Electronic data processing in geographical research. *Professional Geographer* 14:1–4.1962.

[3.] Monmonier, M. Strategies for visualization of geographic time-series data. *Cartographica* 27:30–45.1990.

[4.] Peuquet, D. J. and E. Wentz. An approach for time based analysis of spatio- temporal data. Sixth International Symposium on Spatial Data Handling, Edinburgh, Scotland, *International Geographical Union:* 489–504.1994.

[5.] Bell, S. B. M., B. M. Diaz, F. C Holroyd, and M. J. Jackson. Spatially referenced methods of processing raster and vector data. *Image and Vision Computing* 1:211–20.1983.

[6.] Brunner, R., K. Ramaiyer, A. Szalay, A. Connolly and R. Lupton. An object-oriented approach to astronomical databases. Proc. 4th Ann. Conf. Astronomical Data Analysis Software and Systems, Baltimore. 1994.

Questions

1. What are the processing tools of a GIS? How are they manifested in a GIS?
2. Give a block diagram describing the sub-process of geo-processing tools transferring spatial input to output in a GIS.
3. What are the main categories of geo-processing analysis?
4. Enlist the important processing that is carried out in a GIS.
5. What are the different types of height in GIS and explain their relation?
6. Derive the formulae for computing the area of a terrain surface given in the form of a closed chain of segments.
7. Derive the formulae for computing the volume of a terrain surface expressed in the form of contours.
8. Derive the formulae for computing the surface distance between two points on the earth surface?
9. What is the azimuth and the elevation of a celestial body?

5

3D Processing and Analysis

3D terrain visualization is a prime function in many applications and information systems such as GIS, virtual reality, scientific visualization, and remote sensing. Viewing the terrain surface as it is in the digital display has enthused many researchers and user communities due to various reasons. To name a few, a battle commander needs to view a fly through model of the red (enemy) land to have a general appreciation of the terrain, to study the topography and to observe enemy deployment. A pilot may like to have a first-hand experience of flying to a destination along a particular air route before actually commencing the flight.

One can obtain a general appreciation of the terrain by visualizing the 3D terrain with different information layers laid on top of a model e.g. a satellite image, a topographic map or a shaded relief map. Further, different temporal conditions such as daylight, night time, fog, cloud and other meteorological conditions can be simulated on the terrain for realistic visualization of the terrain. Hence of late, 3D terrain viewing and analysis or simulation of a fly through along a terrain has become an integral part of any GIS software.

Terrain surfaces are usually modeled in GIS using grid of raster cells or triangular irregular networks (TINs). The derivatives discussed in this chapter may be calculated using either of the models, but emphasis is given more on raster maps. The most commonly calculated surface derivatives are slope and aspect. Because slope estimates in coordinate systems using latitude and longitude are not interpretable, the derivatives are measured using ground distance units like meters, feet, or miles.

5.1 What is Digital Terrain Modeling?

In general the first question, that comes to the mind of a beginner is: 'What is digital terrain modeling?'

A terrain model is a scaled down visual representation of the terrain surface as a solid model or a computer model displayed in a digital display environment. Creation of a terrain model involves the collection of terrain data, organization of data and finally representation of the data on computer screen or through a solid model. A digital representation of the terrain is often called a digital terrain model (DTM) or a digital elevation model (DEM) and the data is called digital terrain elevation data (DTED) (US Geological Survey 1987).

In a DTM the representation of the terrain is done in the form of a grid of height values which corresponds to the average height of the patch of earth under consideration from the mean sea level. This grid of height values or matrix of values divides the patch of earth under consideration into a matrix of square elements. The average height of the square elements of the earth is assigned the height value of the grid point as depicted in Fig. 5.1. Obtaining the height at each and every point of the grid and associating them with the patch of earth is done using various methods e.g. from ortho-photos, from satellite images, from shaded relief maps, from surveyed spot heights, trig. heights and contours. Different formats of elevation models can be obtained depending on the way height values are organized. The most popular elevation models are DTED, DEM and triangular irregular network (TIN) (Cavendish 1974). Among the DEM models, there are many popular data formats that are in practice e.g. USGS DEM/DTED, National Marine Engineers Association (NMEA) format etc. In general, the data files are organized into two parts: (1) the header and (2) the body. The header contains the meta-data regarding the actual height matrix and the body constitutes of the height profile as a regular grid.

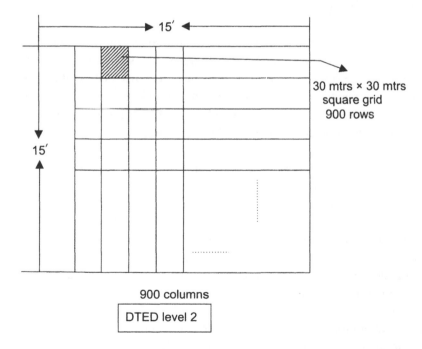

Fig. 5.1 DTED matrix of 15' × 15' area

The evolving spatial modeling techniques and algorithms combined with easy availability of DEM data has made applications of DEM popular. Different applications of DEM in both military and civilian applications are increasing day by day. The importance of DEM can be emphasized from emerging applications such as fly through and walk through simulation, scientific visualization, study of terrain change, etc. Here are some reasons why digital elevation models are important. This is a candidate list and is not exhaustive.

1. DEM combines the following data and information to give a comprehensive visualization of the terrain as it is: the current state of a landscape typically captured in a GIS i.e. the digitized contours, trig. heights (triangle heights), the spot heights surveyed as benchmark heights of the terrain surface and height of the earth surface obtained from GPS.

2. It helps to visualize and plan the future use of the terrain surface and schedules of human activities such as establishment of new human settlements, vital installations, landscape modeling and possibility of land slides and disasters.

3. DEM helps to understand and gain knowledge about how natural processes can get affected because of changing terrain.

4. It helps to produce time-series data regarding the potential future of the landscape. Fly through and walk through simulations can be conducted for visualization of unknown territories where direct human reconnaissance is not advisable or will risk human life.

5.2 Generation of DEM

The choice of sources of data and capturing techniques used for collection of terrain data are critical for the quality of the resulting DTM. DTM data mainly consists of observations of terrain elevation, shape of the terrain surface (i.e. structural features such as drainage channels, ridges, peaks and other terrain surface discontinuity) and spot heights or trig. heights obtained from ground survey (Cole et al. 1990; Zhang and Montgometry 1994). DEM data is directly or indirectly obtained from the following sources

1. Ground survey
2. Photogrammetric data capture (manual, semi-automatic, automatic mode)
3. Digitized cartographic data sources etc.

In addition to the above traditional sources, different sources are used to obtain DEM data for different terrains. Some of the popular techniques of DEM data collection are RADAR (radio detection and ranging) used for mountainous terrain, LASER (light amplification by stimulated emission of radiation) altimetry used by UAV and airborne platforms, LiDAR (light detection and ranging) for mapping city profiles and SONAR (sound navigation and ranging) for aquatic terrain. These techniques help in measurement of undulation in terrain surface using stereoscopy.

DEM data are obtained after processing satellite scans or UAV (unmanned aerial vehicles which help in aerial surveys) scanned images, interpolated heights from ground survey contours, spot heights and trig. heights. Aerial survey through photogrammetry technique is a mechanism of ortho-photo generation and aero-triangulation from which height of ground objects are computed. Of late, ground survey using GPS and DGPS gives the nascent profile of the earth surface in terms of location and height from MSL.

Figure 5.1(a) depicts how an aerial photo of an undulated earth surface is captured using an airplane. The photograph obtained through this sort of an aerial platform has a disadvantage in that it does not capture the height of the terrain. But there will be a shift of objects in the photograph, which has considerable height above mean sea level due to relief of the terrain. This is known as displacement due to relief. This phenomenon of relief displacement in ortho-photo is exploited to measure the height of the object on earth. To obtain the height of the ground, two oblique photographs of the same area is taken from two different perspective points using the same camera and from the same height as depicted in Fig. 5.1 (a) and (b).

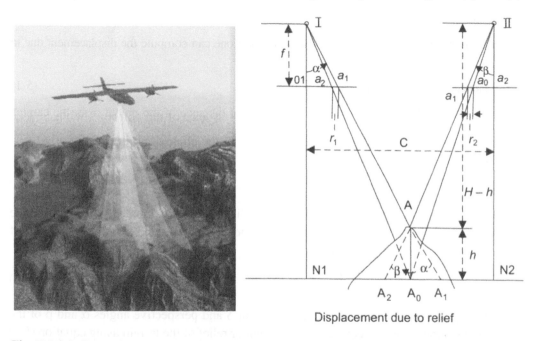

Displacement due to relief

Fig. 5.1 (a) Capturing an aerial photograph of an undulated terrain surface; **(b)** Computing displacement due to relief of ground from the aerial stereo-photo)

These are called stereo pair of photographs. Because of the relief of the ground, although the photographs are taken from the same height, the object will appear to have been displaced from its original position. The height of the terrain object is then computed using this relative shift observed from the pair of stereo-photos (equation (5.1)).

In Fig. 5.1(b) vertical photographs of the irregular terrain surface is taken from two different perspective points I and II with the same camera of focal length f at an altitude of $(H - h)$ above the point A. Rays from A toward the perspective center I pierce the focal plane at point a_1. Had point A been at an elevation $h = 0$, that is, either at sea level or in some other reference plane, the rays would have been reflected from point A_0 and would have pierced the focal plane of the camera at I at point a_0. The distance $r_1 = a_1 a_0$ on the photograph I is the displacement of the image point of A due to relief h above the datum surface. Similarly on photograph II the displacement $r_2 = a_2a_0$ due to relief is recorded from the same camera. From Fig. 5.1(b) using geometry one can establish

$$O_1a_1/f = \tan \alpha$$

where O_1a_1 is the distance on the photograph from the principal point to the picture location of A and α is the angle of the ray IA from the vertical. From Fig. 5.1(b) using similarity of triangles one can establish

$$\frac{r_1}{A_0A_1} = \frac{f}{H}$$

Substituting $A_0A_1 = h \tan \alpha$ in the above equation one can compute the displacement due to relief r_1

Hence $r_1 = h f \tan\alpha /H$ (5.1)

The scale of the photograph is the ratio of the actual distance of the object from the camera to its focal length. Hence substituting scale $S = \dfrac{H}{f}$ of the photograph in equation (5.1)

$$r_1 = \frac{h}{S} \tan$$ (5.2)

Using the same principle in stereo photo II, the relief r_2 can be computed. The height of the terrain h can be computed more accurately using the differential relief obtained by the pair of ortho-photos as given by equation (5.3)

$$r = r_2 - r_1 = \frac{h(\tan \alpha - \tan \beta)}{S}$$ (5.3)

Therefore, given relief displacement r, scale factor S and perspective angles α and β of the imaging instrument one can compute h, the height or relief of the terrain using equation (5.3)

5.3 Collection of DEM

Various survey and data collecting agencies over the years have collected digital elevation data using a number of different production strategies such as manual profiling from photogrammetric stereo-models; stereo-model digitizing of contours; digitizing topographic map contour plates; converting hypsographic and hydrographic tagged vector files; and performing auto-correlation via automated photogrammetric systems. Of these techniques, the derivation of DEMs from vector hypsographic and hydrographic data produces the most accurate model, and is the preferred method in use.

One important source of DTED data is the Internet. The Shuttle Radar Topography Mission (SRTM) successfully collected interferometric synthetic aperture radar (IFSAR) data over 80 percent of the landmass of the earth between 60 degrees north and 56 degrees south latitudes in February 2000. The mission was co-sponsored by the National Aeronautics and Space Administration (NASA) and the National Geospatial Intelligence Agency (NGA) of USA. NASA's Jet Propulsion Laboratory (JPL) performed preliminary processing of SRTM data and forwarded partially finished data directly to NGA for their contractors to finish it. Subsequently these data were made available to different US agencies and are now available through the Internet for the general public.

5.4 Input Domain of DEM

Visualization of DEM varies according to the data contents, visualization techniques used and requirement or purpose of the model. Generally the input domain of a DEM constitutes the following components.

1. The surveyed contours (iso-heights)
2. The spot heights or heights obtained from GPS/DGPS
3. The triangle heights or trig. heights
4. The randomly sampled heights from MSL for a set of surveyed points of the earth surface
5. Grid of height values representing the statistical height of a patch of earth
6. The satellite image of the area or image taken from UAV
7. The almanac data of the area with respect to time
8. The layer-wise vector data

5.5 Process Flow of 3D Analysis

The block diagram in Fig. 5.2 depicts the process flow of how 3D processing is carried out on the digital terrain elevation data (DTED) obtained from various sources. Steps 1–4 explain the process.

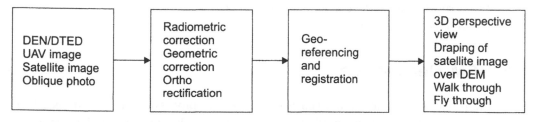

Fig. 5.2 Process flow of 3D processing

STEP 1   The height data are obtained through various sources such as surveyed spot heights, triangle heights, surveyed contours and heights obtained from GPS and DGPS.

STEP 2   A regular grid of known resolution, i.e. a grid of known rows, columns and *inter se* distance between rows and columns is imposed on the above data so as to obtain a uniform grid of height values from these scattered data using interpolation. Usually a bilinear interpolation is used to obtain a uniform array of heights from the above data.

Sometimes an irregular grid e.g. Delaunay triangulation is constructed from the scattered data to obtain a unique triangular irregular network (TIN) of height values.

STEP 3   Finally the TIN or the grid is geo-coded and geo-referenced i.e. the upper left and lower right corner boundary of the data is assigned the actual geographical coordinates (latitude, longitude) pertaining to the ground. This defines the physical boundary of the data representing the terrain profile. Thus the coordinate readout will read the true (latitude, longitude and height) from MSL or (easting, northing and altitude) from MSL at every point.

Finally the following visual analysis can be performed on the DEM.

STEP 4   3D perspective transformation is applied to the data with respect to the viewer's eye point to visualize a 3D perspective of the terrain.

A shaded relief map is generated for different azimuths and elevation of the sun depicting the relative height of the terrain due to the shade generated from the sunlight.

A colour code is applied to different ranges of heights to generate a colour-coded view of the terrain profile. A terrain irregular view and a wire frame model gives different views of terrain undulation.

Draping of the satellite image or scanned photo of the area is done over the DEM to give a realistic view of the terrain. Simulation of daylight, twilight or almanac condition over the terrain gives a realistic visualization of the terrain surface.

The followings can be computed using DEM data. These are known as the DEM derivatives.

1. Slope at each point on the terrain
2. Rate of change of slope
3. Direction of the slope from the north
4. The height from mean sea level
5. The minimum and maximum height of a particular area
6. The profile of the land; cross-section of a designated path (path profile)
7. Linear distance taking into account the undulated surface between two points
8. The perimeter and surface area of a closed area
9. The volume of a piece of land

5.6 DEM Derivatives

Sets of information that can be derived from DEM after processing are known as DEM derivatives (Guth 1995; Burrough 1986). Depending upon different processing methods, different results are obtained from the DEM data. Some significant and commonly used information derived from DTM are the height from MSL, the slope, aspect and curvature. These derivatives have already been discussed in Chapter 4 where terrain surface was modeled as a discrete matrix of heights. In this chapter, the height at a terrain location will be modeled as a continuous function of the location, i.e. height is a function given by $h = f(x, y)$ which perfectly models a raster image of a terrain surface.

5.6.1 Slope

Slope measures the steepness of the surface at any particular location. It is often measured in degrees or in percent rise. A flat region has a slope equal to zero. The steeper the surface, the higher is the slope. Slope is an important variable for many analyses; here are just a few:

- To generate drainage patterns on digital elevation models (DEMs)
- For environmental modeling
- For selection of suitable site for staging, installation of equipments etc.
- For generation of cross-country movement models

A map obtained by the derivative of DEM data using neighbourhood operations is known as a slope map. A slope map gives the rate of change of the height of the cell under consideration with respect to its adjacent cells. It gives a measure of the steepness of an area of the earth's surface. Slope is usually measured in degrees or percent. The slope value assigned to each cell reflects the overall slope based on the relationship between that cell and its neighbours.

5.6.2 Aspect

Aspect measures the direction of the steepest slope for a location on the surface. It is usually measured in degrees, where 0 degrees is due north, 90 degrees is due east, 180 degrees is due south, and 270 degrees is due west (Fig. 5.3). Note that zero and 360 degrees are equivalent.

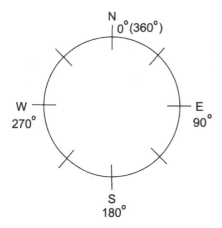

Fig. 5.3 Aspect: The orientation of slope measured in degrees from north

A map where each cell is assigned a value of its orientation in the form of an angle is known as an aspect map. An aspect map tells us the direction to which the surface cells are oriented with respect to the terrain normal. Generating and interpreting maps of aspect can be tricky, since the scale 'wraps' around from 359 to 0 degrees.

5.6.3 Curvature Map

Curvature gives the rate of change of slope. In a curvature map each cell carries the value of rate of change of the slope i.e. curvature of the landform. Geomorphology uses curvature maps for landform analysis (concave/convex) to study the ageing of terrain surface or for change detection of the terrain surface. The age of the terrain surface can also be gauged from the curvature.

If the height is given, the slope, aspect and curvature of a patch of terrain can be computed using the following formulae. In this, the terrain height is given modeled as a continuous function of the location (x, y) i.e. $h = f(x, y)$.

$$\text{Slope} = \sqrt{\left(\frac{\partial h}{\partial x}\right)^2 + \left(\frac{\partial h}{\partial y}\right)^2} \tag{5.1}$$

$$\text{Aspect} = \tan^{-1}\left(\frac{\dfrac{\partial h}{\partial x}}{\dfrac{\partial h}{\partial y}}\right) \tag{5.2}$$

The second order derivative of the DTM grid (i.e. the first derivative of the slope) gives the curvature of the DTM. The curvature describes the convex, concave or plane nature of the surface.

$$\text{Curvature} = \sqrt{\left(\frac{\partial^2 h}{\partial y^2}\right)^2 + \left(\frac{\partial^2 h}{\partial y^2}\right)^2} \tag{5.3}$$

In case of a DEM, the height of the terrain surface is given as a matrix of values with each value of the matrix element representing the height of a terrain cell (Fig. 5.4). In other words, in DEM, height is not a continuous function but a discrete value of a patch of earth.

A	B	C
D	E	F
G	H	I

Fig. 5.4 Grid of DTM with cell designated alphabetically

Hence the computation for slope, aspect and curvature is done differently (equation (5.4)). In Fig. 5.4, let us consider Z_a to be the elevation for cell A, Z_b the elevation for cell B, and so on. Then, the algorithm for finding the slope in cell E is given in the steps below.

STEP 1   Calculate the east–west gradient (dz/dx) and north–south gradient (dz/dy) for cell E:

$$\frac{dz}{dx} = \frac{(z_a + 2z_d + z_g) - (z_c + 2z_f + z_i)}{8 \times \text{cell resolution}}$$

$$\frac{dz}{dy} = \frac{(z_a + 2z_b + z_c) - (z_g + 2z_h + z_i)}{8 \times \text{cell resolution}}$$

STEP 2   Calculate the slope at E from the partial gradients using equation (5.4)

$$\% \text{ of slope at E} = \sqrt{\left(\frac{dz}{dx}\right)^2 + \left(\frac{dz}{dy}\right)^2} \tag{5.4}$$

Note that dz/dx and dz/dy are equivalent to rise/east–west run and rise/north–south run, respectively. The aspect and curvature of a DEM can be computed using the values of east–west gradient and north–south gradient substituted in equations (5.2) and (5.3) respectively.

5.7 Processing and Visualizations of 3D Data

There are various ways of visualizing the ⋅ surface of the terrain using GIS—3D perspective view, an orthogonal view, a perspective view with a colour-coded relief map draped on the undulated terrain and a satellite image draped on top of the terrain surface. All these give a static view of the undulated terrain. On the other hand, simulations such as a fly through the terrain at a specified height and a walk through the terrain give a dynamic visualization of the terrain surface. Figures 5.6–5.11 show a specimen of each of these types of visualization possible through a modern GIS system.

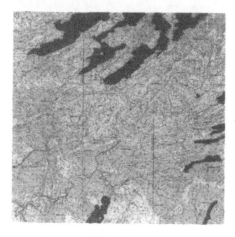

Fig. 5.5
Raster image/scanned image
of the map

Fig. 5.6
3D perspective view of the DEM with
raster image and vector layer draped on it

Fig. 5.7
Shaded relief view of the map

Fig. 5.8
Colour-coded elevation map

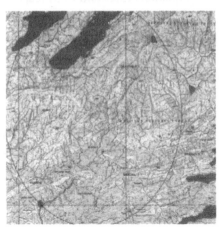

Fig. 5.9
Computing the minimum and maximum
height of a circular area

Fig. 5.10
Line of sight in the view cone
(red line: invisible; green line: visible)

Fig. 5.11
Fly through visualization over the area with
raster image overlaid

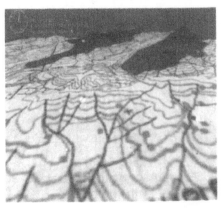

(Also see Plates 1, 2 and 3, for Figs 5.5–5.11)

5.8 Applications of DEM

DEM provides a forum for formalized thinking about how the landscape is working (Zhang et al. 1994). Users are already making decisions based on conceptual models prepared using DEM. It helps users to formally analyze, improve, and combine formalized conceptual models and geo-spatial data from various sensors and sources and identify the relative importance of the collection of different data.

Various applications use DEM for generation of three-dimensional graphics displaying terrain slope, aspect (direction of slope), and terrain profiles between selected points. DEMs have been mostly used in combination with digital raster graphics (DRGs), digital line graphs (DLGs), and digital ortho-photo quadrangles (DOQs) to enhance the visual information. They are also used for data extraction, revision purposes and to create aesthetically pleasing and dramatic hybrid digital images. Non-graphic applications such as terrain modeling and gravity data modeling used for searching for energy resources, calculating the volume of water reservoirs, and determining landslide probability have also been developed using DEM extensively.

Oil and gas

DEM is used for locating oil wells and planning routes and pipelines for transport of oil and gas. In telecommunications, DTED data is used to plan installation of telecom infrastructures such as transreceivers and antennas for wireless networks supporting propagation models integrated with clutter models. In land use study, environment engineering, study of erosion and forestry, hydrology and drainage, DTED is used extensively.

Slope derivatives of DEM can be used to generate attractive hill-shaded maps with different sun-angles. One such sample output of a GIS using DEM data, computing the shades generated due to slope, is depicted in Fig. 5.12.

Batalik: South view

Fig. 5.12 Hill shade depicting flow patterns

References

[1.] Paul A. Longley, Michael F. Goodchild, David J. Maguire and David W. Rhind (eds). *Geographical Information Systems (Principles & Technical Issues) vol.1, vol. 2, 2nd ed.* John Wiley & Sons. 1999.

[2.] Paul A. Longley, Michael F. Goodchild, David J. Maguire and David W. Rhind. *Geographic Information System & Science.* John Wiley & Sons. 2001.

[3.] Burrough, P.A. *Principles of Geographical Information Systems for Land Resources Assessment.* Chapter 3. Oxford: Clarendon Press. 1986.

[4.] Guth, P.L. Slope and aspect calculations on gridded digital elevation models: Examples from a geo-morphometric toolbox for personal computers. *Zeitschrift für Geomorphologie* 101: 31–52. 1995.

[5.] Digital Elevation Models: Data Users Guide 5. USGS, US Dept. Interior, Reston, VA. 1987.

[6.] Weibel, R. and M. Heller. Digital terrain modeling. In D.J. Maguire, M.F. Goodchild and D.W. Rhind (eds). *Geographical Information Systems, Principles & Applications.* London: Longman.1991. 269–297.

[7.] Ackermann, F. Experimental investigation into the accuracy of contouring from DTM. *Photogrammatic Engineering and Remote Sensing* 44:1537–48.1978.

[8.] Cole, G., S. MacInnes and J. Miller. Conversion of contoured topography to digital terrain data. *Computers and Geosciences* 16:101–9.1990.

[9.] Bolstad, P.V. and T. Stowe. An evaluation of DEM accuracy: Elevation, slope, and aspect. *Photogrammetric Engineering and Remote Sensing* 60: 1327–32.1994.

[10.] Tribe, A. Automated recognition of valley features for digital elevation models and a new method toward their resolution. *Earth Surface Processes and Landforms* 17:437–54. 1991.

[11.] Zhang, W. and D.R. Montgometry. Digital elevation model grid size, landscape representation, and hydrologic simulation. *Water Resources Research* 30: 1019–28. 1994.

[12.] Ackermann, F. Experimental investigation into the accuracy of contouring from DTM. *Photogrammertic Engineering and Remote Sensing* 44:1537–48. 1978.

[13.] Bolstad, P.V. and T. Stowe. An evaluation of DEM accuracy: elevation, slope, and aspect. *Photogrammetric Engineering and Remote Sensing* 60: 1327–32. 1994

[14.] Cole. G., S. MacInnes and J. Miller. Conversion of contoured topography to digital terrain data. *Computers and Geosciences* 16:101–9. 1990.

[15.] Cavendish, J.C. Automatic triangulation of arbitrary planar domains for the finite element methods. *International Journal for Numerical Methods in Engineering* 8: 679–696. 1974.

Questions

1. What is DTM (digital terrain modeling)?
2. How is a DEM (digital elevation model) generated?
3. How do we compute relief (r) of an earth surface from two stereo-photographs taken at different perspective angles α, β respectively, at a height (h) from the surface?
4. What is the input domain of a DEM?
5. Give a block diagram describing the flow of processing while generating DEM data in GIS.
6. What are DEM derivatives? How are they useful?
7. Enlist some prominent applications of DEM.

6

Coordinate Systems and Referencing Earth Objects

The concept of spatial and temporal positioning of earth features in a physical container such as a paper map, clay or sand model is very fundamental to GIS. It calls for a relative positioning of the features with respect to each other and with respect to an overall frame of reference. The concept of frame of reference in modeling has been extended directly to digital modeling of spatio-temporal data in GIS. Hence, spatial and temporal positioning of earth features needs a reference frame. Reference frames which are otherwise known as coordinate systems, imparts a referencing mechanism to cartographic data. There are different types of reference frames in use depending on the way they impart relative positioning of objects from the whole to a part. This chapter defines coordinate systems and explains why they are required. The different heads under which the coordinate systems are categorized depending upon the unique property they share is discussed. Coordinate transform is extensively used in cartography in general and GIS in particular. Popular coordinate systems used in cartography such as earth centered—earth fixed (ECEF), Universal Transverse Mercator (UTM) and Military Grid Referencing System (MGRS) are discussed in the end.

6.1 What is a Coordinate System?

The overall frame of reference, which imparts the spatial ordering among earth features, is known as the 'universal frame of reference' or 'world coordinate system'. This is essential for modeling of earth, map-making, spatial search, query modeling and analysis. A frame of reference is necessary to define the position of geographical entities in terms of points, lines or polygons to be placed with reference to each other and with respect to the container. A common frame of reference, which provides this, is called the 'local coordinate system' or 'map coordinate system'. The coordinate system and map projection are interlinked; in fact, coordinate transformation is the means of map projection (Maling 1992).

Generally, the coordinate system used while capturing data is different from that used to represent them on a paper map or a digital display board such as a computer display or data wall. Usually the coordinate system used while capturing the data surveying is known as the 'relative frame of reference'. The coordinate system used to capture the data using digitization is an 'intermediate frame of reference' relative to the device and is known as a 'device coordinate system'; and the data represented on a computer monitor or data wall uses a relative coordinate system defined by the display system often known as the 'display coordinate system' or 'view port coordinate system'.

From the foregoing discussions it is clear that there is a global frame of reference with respect to the globe or earth, an intermediate reference system with respect to the data-capturing device such as digitizer/ scanner/ camera and a coordinate system with respect to the display device or view port. The different reference frames involved in modeling, capturing and display of spatial data is depicted in Fig. 6.1.

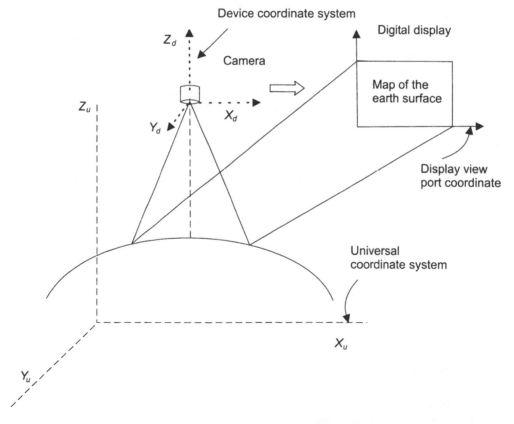

Fig. 6.1 Universal, device and display coordinate system

As spatial data modeled in the world coordinate system needs to be represented in different coordinate systems for capturing and display, there is a frequent need to transform spatial data from one frame of reference to another frame of reference for processing, analysis and visualization. These transformations are known as 'coordinate transformations'. Cartographers have established well-defined mathematical mechanisms to perform these transformations. This mathematical basis has undergone changes and optimizations to adapt to the digital computing environment. Every GIS has options to transform cartographic data in one coordinate system to another. The set up or workbench equivalent of contemporary cartographic systems has been directly replaced by modern GIS workspace, which associates a primary coordinate system to the digital display system in referencing the spatial objects.

6.2 Definition of a Coordinate System

To establish and explain the inter-relationship between different types of coordinate systems it is necessary to have a formal definition of it, both qualitative and mathematical (Peter H. Dana's *The Geographer's Craft Project*). A coordinate system is defined or characterized by its origin, unit of measurement and reference axis. Some of the popular coordinate systems, which are in use in cartography, are given in Table 6.1. Mathematically, a coordinate system is defined by eight parameters viz.

1. A set of axis or reference frame (X, Y, Z)
2. The point of intersection of the axis or the origin i.e. (X_0, Y_0, Z_0)
3. The shift and orientation of the reference frame with respect to an absolute frame of reference, (α, θ, ϕ)

If the axes are mutually perpendicular to each other at the point of intersection then the frame of reference is called an *orthogonal frame of reference*; otherwise it is a non-orthogonal frame of reference. Also, depending upon the number of axes, the coordinate system is characterized as one-dimensional (1-axis), 2-dimensional (2-axes) or 3- dimensional (3-axes) etc.

A coordinate system can be simply defined as a set of rules that specifies how coordinates are assigned to locations of features or objects.

6.3 Why is a Coordinate System Required?

Given below is a list of reasons why coordinate systems are essential for cartography (Maguire et al. 1991).

1. Coordinate systems impart spatial ordering of the features with respect to a common frame of reference in terms of location, distance, orientation or direction from a fixed frame of reference.
2. Coordinate systems help organizing the spatial data in the data capturing system, data producing system and storage systems to record, generate and store digital data. In fact primary and secondary coordinate systems are treated as the key fields of the records describing the spatial objects in a relational database.

3. Coordinates are a convenient method of recording position in 2D plane or 3D space.
4. Well-established principles of coordinate geometry can be applied on the spatial data for measurement and quantification of fundamental geometric measures such as position, distance, perimeter and area etc. of the spatial features.

Hence coordinate systems are crucial for recording spatial data while surveying, digitization, aerial photography or capturing of satellite images. A special branch of cartography known as computational geography is evolving which combines mathematical computing with the algorithms to evolve efficient programs to analyze and visualize spatial data in computers.

6.4 Categorization of Coordinate Systems

There are a number of coordinate systems that have been used since ages for their mathematical uniqueness to model a particular kind of spatial feature. To abstract and highlight a specific characteristic earth surface, specific coordinate systems have been devised. The classification of various coordinate systems used in cartography is shown in Table 6.1 and Fig. 6.2.

To impart spatial ordering to geographical features, the frame of reference or coordinate system has to be fixed with the conceptual model of the earth and the positions of all features have to be measured with reference to the coordinate system. Two widely used two-dimensional coordinate systems are the Cartesian coordinate system and the polar coordinate system. The popular three-dimensional or spatial coordinate systems used to locate data on the surface of the earth are *spatial geographic* coordinates (φ, λ, h) (Fig. 6.3) and *geocentric* coordinates (x, y, z). These are the two coordinate systems that are generally in use for earth modeling. The spatial geocentric coordinate system is also known as the 'earth centered—earth fixed' (ECEF) system (Fig. 6.4), where the origin of the system is the center of mass of the earth (Peter H Dana's website).

Table 6.1 Coordinate systems used in cartography

	Cartesian / Planar	**Spherical/Spatial**
Geographic	Planar geographic coordinates (φ, λ)	Spatial geographic coordinates (φ, λ, h)
Cartesian	Planar Cartesian coordinates (x, y) or easting and northing (E, N)	Spatial geocentric coordinates (x, y, z) or easting, northing and altitude (E, N, A)
Polar	Two-dimensional polar coordinates	

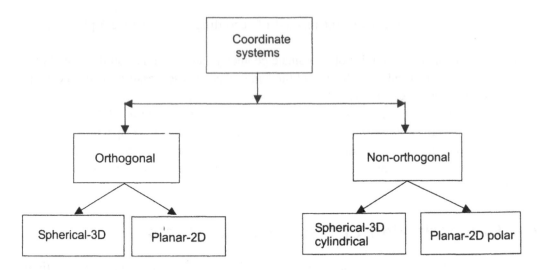

Fig. 6.2 Category of coordinate systems

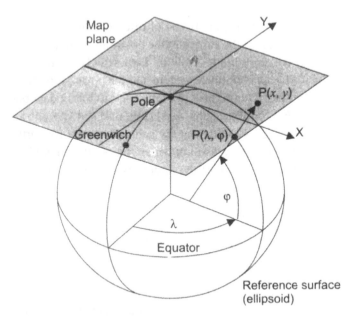

Fig. 6.3 The spatial geographic coordinates (φ, λ, h)

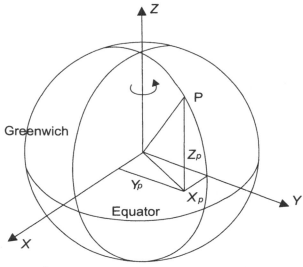

Spatial geocentric coordinates (φ, λ, *h*)

Fig. 6.4 The spatial Cartesian coordinates (ECEF)

6.5 Spatial coordinate systems

6.5.1 Geographic coordinates

The most widely used global coordinate system consists of lines of geographic latitude and longitude. Lines of equal latitude are called parallels. They form circles on the surface of an ellipsoid. Lines of equal longitude are called meridians and they form ellipses (meridian ellipses) on an ellipsoid.

The latitude φ of a point P (Fig. 6.5) is the angle between the ellipsoidal normal through P and the equatorial plane. Latitude is zero on the equator φ (= 0°) and increases towards the two poles to maximum values of φ = +90 (90° N) at the North Pole and φ = −90° (90° S) at the South Pole.

The longitude λ is the angle between the meridian ellipse, which passes through Greenwich, and the meridian ellipse containing the point in question. It is measured in the equatorial plane from the meridian of Greenwich λ = 0° either eastwards through λ = + 180° (180° E) or westwards through λ = −180° (180° W).

The latitude and longitude represents the 2D *geographic coordinates* (φ, λ) of a point P with respect to the selected reference surface such as ellipsoid, geoid or spheroid.

Three-dimensional spatial geographic coordinates (φ, λ, *h*) are obtained by introducing ellipsoidal height *h* to the system. The ellipsoidal height of a point is the vertical distance of the point in question above the ellipsoid. It is measured in distance units along the ellipsoidal normal from the point to the ellipsoid surface. The concept can also be extended to a spheroid or geoid as the reference surface.

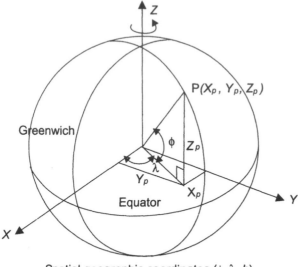

Spatial geographic coordinates (ϕ, λ, h)

Fig. 6.5 Defining the position of a point P using the spatial geographic coordinate system

6.5.2 Geocentric Coordinates (*x, y, z*)

An alternative and often more convenient method of defining a position is by using the spatial Cartesian coordinate system. The system has its origin at the mass center of the earth with the X- and Y-axes in the plane of the equator. The X-axis passes through the meridian of Greenwich, and the Z-axis coincides with the earth's axis of rotation. The three axes are mutually orthogonal and form a right-handed system. This is also known as the 'earth centered—earth fixed' (ECEF) coordinate system.

6.6 Coordinate Transformation

For the purpose of cartography, it is necessary to transform coordinates from a 3D to a 2D planar coordinate system. This requires transformation of the spherical coordinates represented by (latitude, longitude, height) to planar coordinates (*x, y, z*). These transformations are called coordinate transformations and are realized through a series of transformation equations which are in turn expressed as a set of mathematical formulae and realized through computer programs.

To produce a map, the curved reference surface of the earth, approximated by an ellipsoid or a spheroid, is transformed to the flat plane of the map by means of a map projection. In other words, each point on the reference surface of the earth with *geographic coordinates* (ϕ,λ) may be transformed to a set of *Cartesian coordinates* (*x, y*) representing positions on the map plane. The block diagram given in Fig. 6.6 depicts the work flow of various coordinate transformations often practiced in GIS. The mathematical treatments of the transformations are given in the following sections. Although there are other coordinate transformations, they are not practiced frequently in cartography and hence are not included in this book.

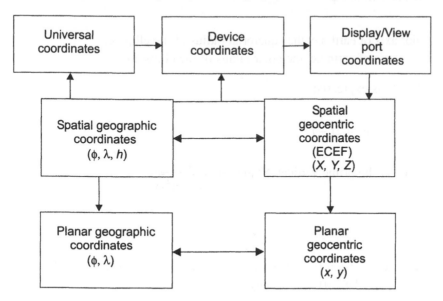

Fig. 6.6 Work flow of various coordinate transformations

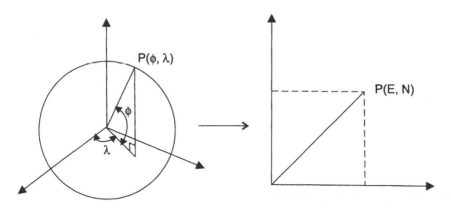

Fig. 6.7 Transformation from spherical coordinates to Cartesian coordinates

6.6.1 Geodetic to Geocentric Transformation

The formulae for conversion of geodetic coordinates (ϕ, λ, h) i.e. (latitude, longitude and height above ellipsoid) to earth centered—earth fixed (ECEF) Cartesian coordinates (x, y, z) are given by (Snyder 1994):

$$x = (N + h) \cos \varphi \cos \lambda$$

$$y = (N + h) \cos \varphi \sin \lambda$$

$$z = ((1 - b^2/a^2) \, N + h) \sin \varphi$$

where,

a = Semi-major axis of earth i.e. the equatorial radius of the ellipsoid

b = Semi-minor axis of earth i.e. the polar radius of the ellipsoid

Flattening of earth is expressed as $f = \dfrac{a-b}{a}$

Eccentricity squared $e^2 = 2f - f^2$

Radius of curvature at the prime vertical $N(\varphi) = \dfrac{a}{\sqrt{1 - e^2 \sin^2 \varphi}}$

6.6.2 Geocentric to Geodetic Transformation

The reverse coordinate conversion formulae for converting (x, y, z) to geodetic coordinates (latitude, longitude, ellipsoid height) are given by:

$$\varphi = \text{arc tan} \left(\frac{z + e'^2 b \sin^3 \theta}{p - e^2 a \cos^3 \theta} \right)$$

$$\lambda = \text{arc tan} \, (y/x)$$

$$h = \left(\frac{p}{\cos \varphi} \right) - N$$

where,

eccentricity squared $e^2 = \dfrac{a^2 - b^2}{a^2}$

θ is an auxiliary quantity = arc tan $\dfrac{za}{pb}$ and $p = \sqrt{x^2 + y^2}$

Often it is necessary to convert the coordinates of the universal frame of reference to a device coordinate system and from a device coordinate system to a display coordinate system and vice versa. These coordinate transformations are carried out for the purpose of data representation and interpretation through various display devices.

6.6.3 Device to View Port Transformation

Computer cartography records the relative coordinates of the features in the local coordinate system which are particular to the device in which the data is captured or viewed e.g. an overhead digitizer or a graphics display screen. The data obtained through the local data acquisition device is independent of the map grid. Hence this data needs to be projected to the map grid coordinates. Also viewing transformation need to be applied while displaying the data in a specific graphic display device. This calls for view port transformation given by:

$$x_i = x_0 + s_x \left(x_{v\,max} - x_i\right)$$

$$y_i = y_0 + s_y \left(y_{v\,max} - y_i\right)$$

$$z_i = z_0 + s_z \left(z_{v\,max} - z_i\right)$$

where, s_x, s_y, and s_z are the scale factors in x, y and z coordinate respectively. x_0, y_0 and z_0 are the translation factors and are usually assigned the minimum values of x, y and z of the view port.

6.7 Shape of Earth

It is essential to understand the shape of earth and give it a mathematical definition before modeling its features. Hence, in many ways modeling the shape of earth is precursor to imparting a coordinate system to the model. The study and modeling of earth is known as geodesy.

The earth is not an exact ellipsoid; deviations from this shape are different for different regions of the earth and are being continually evaluated because it is being dynamically changing. For map projections, the problem has been confined to selecting constants for the ellipsoidal shape and size of the region under consideration. The reference ellipsoids are used for different regions of the earth. In 1942 the International Union of Geodesy and Geophysics (IUGC) officially defined the shape of the ellipsoid. According to the Union, the equatorial radius (semi-major axis) of earth is 6,378,388 meters. The polar radius of earth is 1/127 less than 6,378,388 meters i.e. 6,356,911.9 meters. This is called the international ellipsoid. The statistics of the ellipsoid reveals the following (Peter H. Dana's website):

Equatorial and polar radius vary by 22 km out of 6400 km, or about 1/300

Mt. Everest being +9 km above mean sea level (MSL) and Mariana's trench –11 km below MSL, the total relief of earth is approximately 20 km

But the geoid relief varies only by about several 100 m due to mass imbalances

Assumption of the earth's surface as a sphere, an ellipse or a geoid and projecting it into different conic sections (plane, cylinder or cone) gives different types of projection, which preserve the topological identity of earth features. Thus, different conic sections and their orientation relative to earth surface give different projections. Geodatic datum defines and models the shape of earth. All these are discussed in Chapter 7 under map projection.

6.7.1 Definition of Datum

Geodetic datum defines the base reference surface level of earth for measurement of coordinates and heights. It depends on the ellipsoid, the earth model and the definition of sea level. Geodetic datum is defined by specifying the location of the coordinate system, orientation of the coordinate system and by the dimensions of the reference surfaces.

Horizontal geodetic datum consists of the latitude, longitude of the initial point (origin) of the coordinate system and an azimuth line from this point to some other triangulation station for its direction. Horizontal datum is used for measuring the position of earth objects and their location in preparation of topographic maps and charts.

Vertical geodetic datum is defined from the geoid surface which most closely approximates the mean sea level (MSL). This is used to measure the height of objects located on and above earth surface uniformly.

6.7.2 Datum/Geodetic Datum

Once the coordinate system is fixed, it is important to model and fit the topographic surface to a reference surface having regular geometry so that the geodetic computations are simple and follow regular mathematics. The topographic surface of earth is highly irregular (Everest and Mariana trench are the extremes). This makes it difficult for any geodetic computation using regular mathematics. To overcome this problem, geodesists have adopted some regular mathematical surfaces, called reference surfaces, which approximates the irregular shape of the earth. There are two such popular models: one approximates the surface as a sphere known as spheroid or geoid, which approximates the global mean sea level; the other model with a biaxial reference ellipse known as ellipsoid (Fig. 6.8) have been evolved as the best-fit candidate often used for geodesic computations.

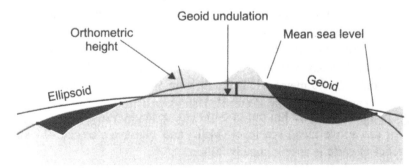

Fig. 6.8 Modeling of earth surface using different datum surface)

The parameters, which define the reference surface, are called the datum of the geometry which defines it. To define earth surface through an ellipsoid and fit it to a regular geometry at least two out of the three parameters enlisted below are necessary:

- a – Equatorial radius, about 6378 km
- b – Polar radius, about 6356 km
- f – flattening, where $f = 1 - b/a$, about 1/298

From the above three parameters, any two define the datum and the third can be derived. Another derived parameter, which defines the geometry of the surface, is eccentricity of the datum surface. Eccentricity is used in most projection formulas. It gives an idea of how rapidly the curvature of the earth is changing or the rate of change of the curvature:

e – Eccentricity; $e^2 = 2f - f^2$

Taking these parameters as the starting data, the imaginary shape of earth is constructed by imposing an imaginary grid which is constructed with the intersection of the Greenwich meridian line (zero longitude) and the equator (zero latitude). The other grid lines are constructed with equal intervals, which are multiples of 6, to obtain a lat.–long. grid.

A reference ellipsoid, as mentioned in the previous section, is known as the geodetic datum. Eight parameters are needed to define a geodetic datum mathematically: two parameters (a, b) to define the dimension of the ellipsoid; three parameters to define the position of the origin; and three parameters to define the three axes with respect to the earth. There are at least a half dozen standard geodetic datum used widely because of their regional accuracy, which best fits or defines the region of earth under consideration. A list of such regional datum is enlisted in Appendix C.

WGS (World Geodetic System) is a geodetic reference system consisting of a set of parameters describing the size and shape of the earth, an earth centered—earth fixed (ECEF) coordinate reference system, the position of a network of points with respect to the center of mass of the earth, and the gravitational model of the earth known as the global geoid.

Every modern GIS supports WGS and many other local datum which are stored as a table so that data obtained in one datum surface can be converted to another datum surface for analysis and visualization.

References

[1.] *Geodesy for Layman*: http://www.nima.mil/geospatial/geospatial.html
This is an html version of the Report TR80-003 published in 1984 by the U. S. Defence Mapping Agency (now the National Imagery and Mapping Agency or NIMA). It explains the basic principles of geodesy, including geodetic datum and satellite geodesy.

[2.] Langley, Richard B. Basic geodesy for GPS. *GPS World* 3:44–49. 1992.

[3.] Maguire, D. J., M. F. Goodchild and D. W. Rhind (eds). *Geographic Information Systems: Principles and Applications*. Harrow, U.K. 1991.

[4.] Maling, D. H. *Coordinate Systems and Map Projections*, 2nd ed. Oxford: Pergamon Press. 1992. Online text available at www.wiley.com/gis and www.wiley.co.uk/gis, Chapter 10, 'Coordinate systems and map projections for GIS'.

[5.] Maling, D. H. *Coordinate Systems and Map Projections*. 2nd ed. Oxford: Pergamon Press, 1992.

[6.] *Map Projection Overview, Coordinate Systems Overview and Geodetic Data, Overview*. The following webpage by Peter H. Dana (The Geographer's Craft Project, Department of Geography, The University of Austin) is an illustration and discussion on these topics. http://www.utexas.edu/depts/grg/gcraft/notes/mapproj/mapproj.html

[7.] *Map Projections – A Working Manual.* U.S. Geological Survey Professional Paper 1395. Washington, D.C.: U.S. Government Printing Office.1987.

[8.] Robinson, A. H., J. L. Morrison., P. C. Muehrcke, A. J. Kimerling and S. C. Guptill. *Elements of Cartography,* 6th ed. New York: John Wiley & Sons Inc. 1995. pp.674.

[9.] Robinson, Arthur H., Randall D. Sale, Joel L. Morrison, and Phillip C. Muehrcke. Elements of Cartography, 5th ed. New York: John Wiley & Sons. 1984.

[10.] Snyder, John P. *Flattening the Earth: Two Thousand Years of Map Projections.* Chicago, IL: University of Chicago Press. 1993.

[11.] Snyder, John P. *Map Projections – A Working Manual.* U.S.G.S. Professional Paper 1395. Washington D.C.: U.S. Government Printing Office, 1987. Reprinted 1989, 1994 with corrections.

Questions

1. What is a coordinate system and why it is required in GIS?
2. What are the different types of coordinate systems used in GIS?
3. How are different coordinate systems used in GIS at different stages of spatial data processing?
4. Derive and explain the formulae transforming geodetic coordinates to Cartesian coordinates in the ECEF system.
5. What is device to view port transformation? Explain with formulae.
6. What are the different shapes of earth used in GIS? How is the shape of earth described in GIS?
7. What is geodetic datum?

7

Map Projection

This chapter deals with the basic concepts of map projection. It answers generic queries such as 'what is map projection?', 'Why is map projection required?', 'Why there are so many projections devised?', 'Which map projections are suitable for a particular application?' etc. How map projections are designed and developed for different regions of earth surface is explained. Classification of different map projections using different development surface, perspective position and position of tangent surface is explained. The mathematical formulae for forward and reverse map projections are derived *ab initio*. Finally the chapter ends with a discussion on how to choose a particular map projection for a specific purpose and for a specific region of the world. All the concepts introduced here are adequately supplemented by illustrations and key notes.

7.1 What is Map Projection?

A map is a two-dimensional piece-wise representation of the earth crust on a paper surface (Snyder 1989; 1994). The shape of earth cannot be equated to any conventional geometric shapes i.e. it cannot be represented or modeled to any standard geometric shape like sphere, ellipse etc. Depicting earth surface on a two-dimensional paper surface without distortion is akin to pasting a peel of orange on the table surface so as to get a uniform and continuous strip on the surface without tearing and stretching the peel, in other words, a near impossible job. But cartographers and scientists have devised a number of mathematical formulae to accomplish the job. For map projections the earth surface is modeled as an ellipsoid, spheroid or geoid for different purposes. Actually, the earth is more nearly an oblate spheroid—an ellipse rotated about its minor axis. Then an appropriate mathematical formula is designed so as to represent the modeled earth surface on to a flat paper surface. These mathematical formulae, which project or translate the earth surface to paper surface with minimum distortion, are known as map projections.

7.2 Why is Map Projection Necessary?

Earth is modeled as a globe for numerous cartographic reasons, but it is not possible to make a globe on a very large scale. Say, if anyone wants to make a globe on a scale of one inch to a mile, the radius of the globe will have to be 330 ft. It is difficult to make and handle such a globe and uncomfortable to carry it to the field for reference. A small globe is rendered useless for referring to a small country or landscape because it distorts the smaller land surfaces and depicts the land surfaces inappropriately. So for practical purposes a globe is least useful or helpful in the field. Moreover it is neither easy to compare in detail different regions of earth over the globe, nor it is convenient to measure distances over it. Hence maps were devised to overcome such difficulties. A map is a two-dimensional representation of a globe drawn on a paper map, which is convenient to fold and carry in the field and easy to compare and locate different parts of earth. Locating a known feature, guiding and navigating from one position to another and comparing two different regions over a map are convenient and easy. Transforming a three-dimensional globe to a two-dimensional paper map is accomplished using map projection. Topographical maps of different scales, atlases and wall maps are prepared using map projections. Thus map projection plays a crucial role in the preparation of different types of maps with different scales, coordinate systems and themes.

7.3 Definition of Map Projection

A map projection is defined as a mathematical function or formula which projects any point (ϕ, λ) on the spherical surface of earth to the two-dimension paper surface (x, y) (Canters et al. 1989)
Hence forward map projections is given by

$$(x, y) = f(\phi, \lambda)$$

Often geographic data obtained in Cartesian coordinates needs to be transformed to spherical coordinates necessitating inverse map projection and is given by

$$(\phi, \lambda) = f(x, y)$$

Thus, the mathematical function which realizes the map projection, essentially projects the 3D world features onto 2D surfaces. Figure 7.1 characterizes the current graphics display technology in map projection. Obviously projections are defined differently for different purposes, which should preserve these different topological properties viz. positional accuracy for point features, distance i.e. relative positional accuracy for line objects, direction of one object with respect to other in the earth surface and area for the polygonal object. Realizing a mathematical function which satisfies all these conditions at the same time is not possible. Hence functions are devised which satisfy some conditions at a time. These functions give rise to maps which are direction and distance preserving but not area preserving. Similarly if they preserve area, then distance between the objects will be inaccurately projected. In fact distance, direction and area are mutually exclusive properties of a paper map. Hence, transforming a 3D datum surface to a 2D map distorts at least one of the fundamental geometrical properties such as shape, area, distance and direction. While preparing different kinds of maps, care is taken to preserve certain properties at the expense of others. Hence different map projections are formulated.

The projection process depends very much upon the shape of earth we take into consideration (McDonnell et al. 1979).

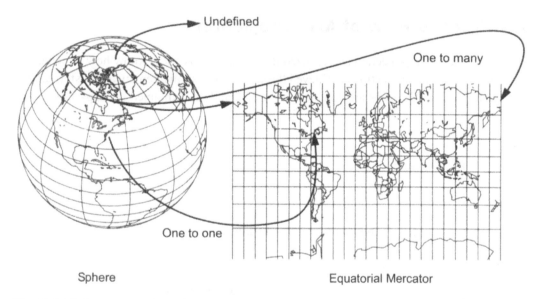

Fig. 7.1 Various scenarios of map projection

Hence map projection function is a 3D-to-2D transformation which preserves the distance, direction, shape and area of the original earth surface. Realizing such a mathematical formula that satisfies such diverse criteria is not possible. Hence different functions, which satisfy different set of criteria, are devised leading to many map projections. Map projections can be done in many different ways. The principal methods transform earth's surface directly to a plane, or to a cylindrical or conical surface. Conceptually a planar, conical or cylindrical surface is *warped* around the earth's surface and unrolled to form the flat surface. The lines marking the latitude and longitude are referred on the plane map as major grid or graticule. The following are some basic definitions of map projection used in different literature.

A projection is a mathematical equation or series of equations, which takes a three-dimensional location on the earth and provides corresponding two-dimensional coordinates to be plotted on a paper or computer screen.

Map projections are treated mathematically as the transformation of geodetic coordinates (ϕ, λ) into rectangular grid coordinates often called easting and northing. This transformation is expressed as a series of equations and implemented as a computer algorithm.

Generally, a globe is considered a true representation of the earth. The globe is divided into various sectors by the lines of latitudes and longitudes. As mentioned before, this network is called 'graticule'. A map projection represents the graticule on a flat surface. Theoretically, map projection might be defined as 'a systematic drawing of parallels of latitude and meridians of longitudes on a plane surface for the whole earth or a part of it on a certain scale so that any point on the earth surface will have correspondence to that on the drawing' (Bugayevskiy 1995).

7.4 Process Flow of Map Projection

Having defined map projection and realized its necessity, it is apt to know how to achieve a map projection. Map projection is achieved through a series of mathematical transformations (Richardus 1972). The steps are depicted in Fig. 7.2 and are explained in Steps 1–4.

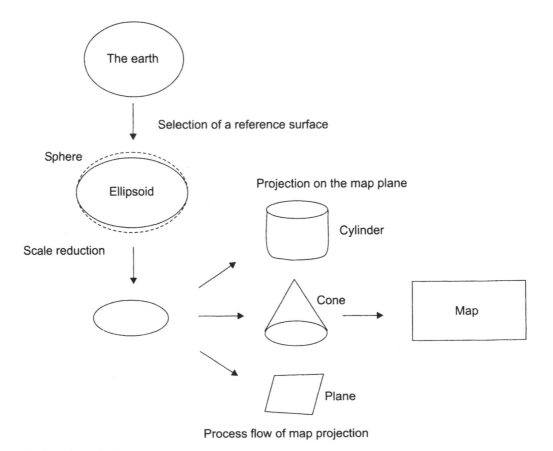

Process flow of map projection

Fig. 7.2 Process flow of map projection

STEP 1 Select a suitable datum surface, which best fits the surface of earth e.g. ellipsoid, geoid or spheroid. A list of different datum surfaces is grouped as a set of values and available in Appendix C

STEP 2 Choose a scale of the earth for mapping to the model globe that can be represented in an appropriate container such as a paper map or clay model etc. Generally, a reduced scale of the actual earth through a representative fraction is chosen.

STEP 3 Keeping in mind the degree of accuracy needed and purpose of map projection, an appropriate projection plane such as cylinder, cone or plane is chosen so as to get a proper projection

STEP 4 With an appropriate perspective position of the observer, the points of representative globe are projected on the projection plane, which is then unwrapped to prepare a map.

7.5 Classification of Map Projection

Keeping the above process flow in mind the following facts about map projection can be enunciated:

1. Theoretically there exists an unlimited number of map projections
2. No map projection is a perfect representation of the actual terrain
3. Map projections can be grouped according to the intermediate projection plane they adopt. e.g. planar, cylindrical or conical
4. Map projections can be grouped according to different perspective positions e.g. gnomonic, stereographic, orthographic etc.
5. Map projections can be grouped according to different types of fundamental geometric quality they preserve e.g. conformal, equidistant etc.
6. Another criterion popularly used for grouping of different class of maps is the representative fraction which decides the scale of the map e.g. 1:50,000, 1:250,000. etc.

Potentially there exits an unlimited number of map projections possessing one property or the other. Class of projections having similar properties can be grouped together for a similar purpose. There is no projection which can be grouped specifically, in a single class. Moreover, if one attempts to obtain a rational classification of map projection, it will be rather difficult to achieve it. Since there are a plethora of map projections available, categorizing them according to a common known property they possess has become absolutely necessary. Also it becomes absolutely necessary to set a guideline for selecting an appropriate map projection for a particular usage. Keeping the above requirements in mind the following classifications are being attempted by various cartographers.

7.5.1 Classification due to Perspective Point

Depicted in Fig. 7.3 are the different types of map projections due to different positions of view point or source of light in the representative globe. This guides the principle of projecting the point on the globe touching the light ray to the projection surface. They are (a) *gnomonic*, where the source of light is placed at the center of the sphere to be projected; (b) *stereographic*, where the light source is at the opposite end of the diameter of the sphere tangent plane; and (c) *orthographic*, where the light source is assumed to be at infinity to the projection plane thus making the rays of light touching the sphere and the projected plane parallel.

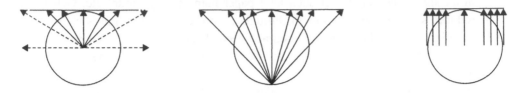

Fig. 7.3 Gnomonic, stereographic and orthographic projection

7.5.2 Classification due to Tangent Plane

Map projections can be categorized according to the position of the tangent surface with respect to the object to be projected (Fig. 7.4). If the tangent surface is perpendicular to the equator then the transformation is known as *normal* projection. If the tangent plane is parallel and touches the pole, it is known as *transverse* projection. If the tangent plane is neither normal nor transverse, then it is known as *oblique* projection.

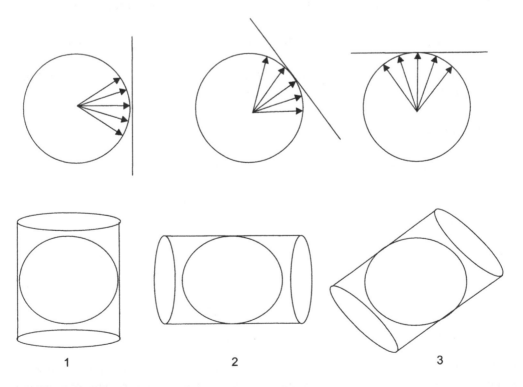

Fig. 7.4 Normal, transverse and oblique map projection

7.5.3 Classification due to Development Surface

One of the important criteria of map projection is the intermediate projection plane or in other words, classification of map projection due to development surface viz. cylinder, cone or plane. The projection surface is important from a cartographer's point of view because it decides the geometry to be applied and the accuracy of the area to be mapped. Diagrams and the mathematical derivation of these projections are given in the next few sections.

Cylindrical Projection

Figure 7.5 gives cross-sectional views of the cylindrical projection. The latitude at ϕ (parallel) on the globe is projected at y on the paper map as a straight horizontal line on the 2D map (Pearson 1977). The transformation equations can be derived by establishing the similarity between the triangles and proportionality of the sides given by

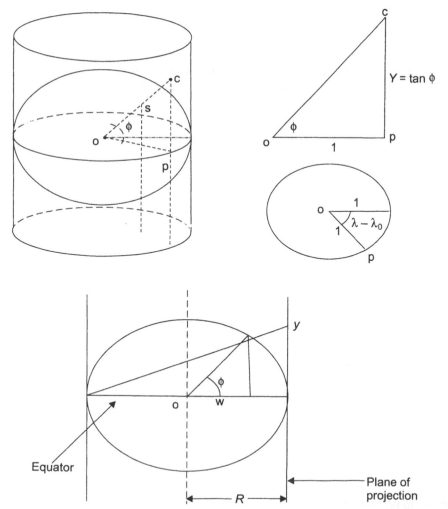

Fig. 7.5 Cross-sectional view of cylindrical map projection

$$\frac{h}{(w+R)} = \frac{y}{(R+R)} \tag{7.1}$$

where, $h = R \sin \phi$ and $w = R \cos \phi$

Substituting the above two values in equation (7.1), one obtains

$$y = \frac{2R \sin \phi}{(1 + \cos \phi)} \tag{7.2}$$

By putting $\theta = \phi/2$ and making use of the trigonometric identity $\sin 2\theta = 2 \sin \theta \cos \theta$ and $\cos 2\theta = 1 - 2 \sin^2\theta$ in equation (7.2), one obtains

$$y = \frac{4R \sin \theta \cos \theta}{1 + 1 - 2 \sin^2 q} = \frac{4R \sin \theta \cos \theta}{2(1 - 2 \sin^2\theta)} = \frac{2R \sin \theta \cos \theta}{\cos^2\theta} = 2R \tan \theta = 2R \tan \frac{\phi}{2} \tag{7.3}$$

Hence, $y = 2R \tan(\frac{\phi}{2})$, and $x = R (\lambda - \lambda_0)$ $\tag{7.4}$

The pair of equations given in (7.4) is the forward cylindrical projection.

Cylindrical map projections are true at the equator, although the distortion increases as one move towards the polar regions. Thus this projection is good for areas in the tropical region of the globe.

Conical Projection

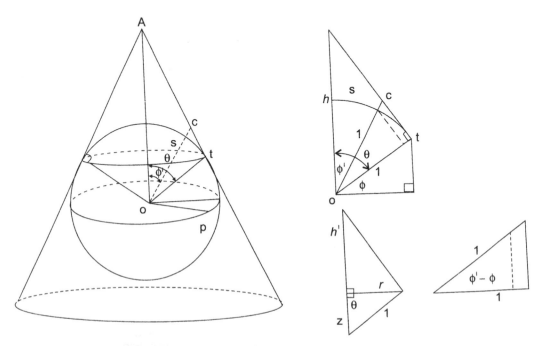

Fig. 7.6 Conical map projection

A conic projection of points on a unit sphere centered at O consists of extending the line OS for each point S until it intersects a cone, wrapping the sphere with apex A, at a point C (Fig. 7.6). The cone which is tangent to the sphere along a circle passing through a point T has height h above O, the angle from the Z-axis at which the cone is tangent is given by equation (7.5).

$$\theta = \sec^{-1} h \tag{7.5}$$

$$r = \sin\theta = \sqrt{\frac{h^2 - 1}{h}} \text{ and}$$

$$z = \cos\theta = \frac{1}{h} \tag{7.6}$$

Equation (7.6) gives the radius of the circle and the height z of the circle above the center of the sphere at O in terms of height h and angle θ.

Let the co-latitude of the point S on the sphere be $\phi' = \pi/2 - \phi$. Length of OC through OS is given by equation (7.7)

$$1 = \sec(\theta - \phi') = \sec^{-1} h - \phi') \tag{7.7}$$

Hence in conic projection one can obtain the Cartesian coordinates (x, y) of a point on the map given the spherical coordinates using the following equations (7.8).

$$x = \sec(\theta - \phi') \cos\phi \sin\left(\frac{\lambda}{\sqrt{\lambda^2 - 1}}\right)$$

$$y = \sec(\theta - \phi') \cos\phi \sin\left(\frac{\lambda}{\sqrt{\lambda^2 - 1}}\right) \tag{7.8}$$

A cone may be imagined to touch the globe of a convenient size along any circle (other than a great circle) but the most useful case will be the normal one in which the apex of the cone will lie vertically above the pole on the earth's axis produced and the surface of the cone will be tangent to the sphere along some parallel of latitude. This parallel is called the 'standard parallel' (Pearson 1977).

If the selected parallel (SP) is nearer the pole, the vertex of the cone will be closer to it and subsequently the angle at the apex will increase proportionately. When the pole itself becomes the selected parallel, the angle of the apex will become 180 degrees, and the surface of the cone will be similar to the tangent plane of the zenithal projection. On the other hand, when the selected parallel is nearer to the equator, the vertex of the cone will be moving farther away from the pole. If the equator is the selected parallel, then the vertex will be at an infinite distance, and the cone will become a cylinder.

Thus the cylindrical and zenithal projections may be regarded as special cases of conical projections. Conics are true along some parallel somewhere between the equator and the pole. Distortion increases away from this standard. Conical projections are good for temperate zone areas.

Azimuthal Projection

A plane tangent to one of the Earth's poles is the basis for polar azimuthal projection. Figure 7.7 illustrates the geometry of azimuthal projection of a point. In an azimuthal map projection, azimuths of all points are shown correctly with respect to the center (Snyder 1987, p. 4).

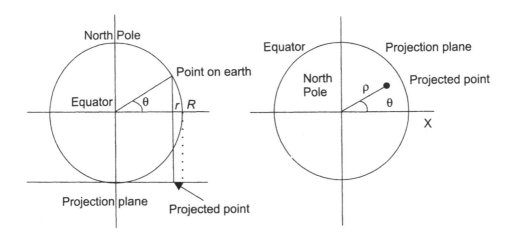

Fig. 7.7 Azimuthal map projection

An azimuthal projection is neither equal-area nor conformal. Assuming a spherical earth model, the forward and inverse transformation equations of azimuthal projection are given by the pair of equations (7.9).

$$x = 2R \times \tan\left(\frac{\pi}{4} - \frac{\phi}{4}\right) x = \sin(\lambda - \lambda_0)$$

$$x = 2R \times \tan\left(\frac{\pi}{4} - \frac{\phi}{4}\right) x = \cos(\lambda - \lambda_0)$$

(7.9)

Inverse transformation equation is given by (7.10)

$$\phi = 2\arctan\left(\frac{\sqrt{(x^2 + y^2)}}{2 \times R}\right) - 90°$$

(7.10)

$$\lambda = \lambda_0 + \arctan\left(\frac{x}{y}\right)$$

In a zenithal projection, a flat paper is supposed to touch the globe at one point and the light may be kept at another point so as to reflect or project the lines of latitude and longitude on the plane. Here the globe is viewed from a point vertically above it, so these are called zenithal projections. They are also called 'azimuthal' because the bearings are all true from the central point.

Depending on the plane's position touching the globe, the zenithal projection is of three sub-classes:

1. Normal or equatorial zenithal (where the plane touches the globe at equator)
2. Polar zenithal (where the plane touches the globe at pole)
3. Oblique zenithal (where the plane touches the globe at any other point)

According to the location of the view point, zenithal projection is grouped into three sub-classes:

1. Gnomonic/central (view point lies at the center of the globe)
2. Stereographic (view point lies at the opposite pole)
3. Orthographic (view point lies at infinity).

Azimuthals are true only at their center point, and generally distortion is worst at the edge of the map. Azimuthal projections are good for polar regions of earth.

7.6 Choice of Map Projection

Out of a theoretically infinite number of map projections available, the cartographer or GIS user has to judiciously choose a map projection, which will satisfy his needs. The factors that lead to the choice of a particular map projection is: (i) the reason and extent of the earth surface to be mapped; (ii) the purpose for which the map will be used e.g. navigation, cadastral mapping, city planning etc; (iii) the fundamental quantity that will be required to be preserved for the application; and (iii) finally the scale factor, accuracy and coordinate system required (Snyder 1987).

Based on the above criterion, the characteristics and the utility of the projection can be ascertained. Table 7.1 suggests classifications based on different criterion and will help a cartographer or a user to zero onto the appropriate type and kind of map projection to be used (Richardus 1972).

Table 7.1 Map projection vs basis of map classification

Sl. No	Basis of classification	Types of map projection
1.	Method of construction	Perspective Non-perspective
2.	Preserved geometrical quantity	Equal area/homolographic Area preserving, shape preserving, bearing (distance and direction) preserving Conformal/orthomorphic
3.	Development surface	Conventional/planar Cylindrical Conical Azimuthal/zenithal

(continues...

Table 7.1 continued)

Sl. No	Basis of classification	Types of map projection
4.	Position of the tangent surface of projection surface	Equatorial Polar Oblique
5.	Position of the view point or light source	Gnomonic Stereographic Orthographic Others

Table 7.2 enlists some examples of projections used for different types of maps. This will give a general idea to the user of the factors that result in a particular choice of map class.

Table 7.2 Map projections used for preparation of different types of maps

Sl No	Type of usage of map	Map projection used
1	Topographical maps	Polyconic projection, Lambert conformal conic (LCC), Stereographic azimuthal, Transverse Mercator, UTM etc
2	Geographical maps (large extent in N–S direction)	LCC projections
3	Cadastral maps (small extent in E–W direction)	Cassini projection, LCC
4	Air and sea navigation maps	Mercator projection (Equatorial regions) Gnomonic projection (Polar regions)
5	World maps in one piece	Lambert's azimuthal equal area Mercator projection Gall's stereographic cylindrical
6	Maps depicting continents in an atlas	Simple conic, Bonne's equal area Lambert's azimuthal equal area LCC
7	For maps of celestial objects e.g. Moon surface, Mars surface	Sinosodial projection

7.7 Projections Used For Different Regions of Earth

To define the spatial reference of a country, one needs information about the projection system, and the values of the projection parameters that are suitable and adopted by the country. The other crucial data that are required for the region to be mapped are

R = Radius of the spherical earth model adopted;

(a, b) = Equatorial and polar radius of the reference ellipsoid modeling that region of earth;

The geodetic datum used; the origin of the coordinate system;

The false easting and false northing of the grid system being used by the country;

The central meridians and standard parallels of the center of projections etc

In general Table 7.3 gives a thumb rule of the type of map projection being used in different regions of the earth.

Table 7.3 Map projection used for different regions of earth

Sl No	Region of earth	Projection type
1	Polar regions	Azimuthal projection
2	Tropical regions (mid-latitude regions)	Conical projections
3	Equatorial regions	Cylindrical projections

7.8 Affine Transformation

When the map data or images are obtained without sufficient data regarding its origin, map boundary or transformation parameters, it becomes difficult to project the map to a known geographical location and obtain proper coordinates of its features. Hence a study of the ground features of the image is carried out and the known features (ground correlation points) are marked. A correspondence is established with these ground correlation points to the respective control points of a known map or image and a polynomial transformation is established to register the unknown map to a known map to obtain the correct coordinates. This process is called affine transformation. Affine transformation plays an important role while registering images obtained from UAV, satellites, scanners or unknown sources to a known image for removing geometric corrections in the image due to sensor motion.

An affine transformation preserves co-linearity (i.e., all points lying on a line initially still lie on a line after transformation) and ratios of distances (e.g., the mid-point of a line segment remains the mid-point even after transformation). In this sense, affine indicates a special class of projective transformations that do not move any object from the affine space to the plane at infinity or conversely. An affine transformation is also called an affinity.

Geometric contraction, expansion, dilation, reflection, rotation, shear, similarity transformations, spiral similarities, and translation are all affine transformations, as are their combinations. In general, an affine transformation is a composition of rotations, translations, dilations, and shears.

While an affine transformation preserves *proportions* on lines, it does not necessarily preserve angles or lengths. Any triangle can be transformed into any other by an affine transformation, so all triangles are affine and, in this sense, affine is a generalization of the congruent and similar properties.

A particular example combining rotation and expansion is the rotation–enlargement transformation

$$\begin{bmatrix} x' \\ y' \end{bmatrix} = s \begin{bmatrix} \cos\alpha & \sin\alpha \\ -\sin\alpha & \cos\alpha \end{bmatrix} \begin{bmatrix} x - x_0 \\ y - y_0 \end{bmatrix}$$

$$= s \begin{bmatrix} \cos\alpha(x - x_0) + \sin\alpha(y - y_0) \\ -\sin\alpha(x - x_0) + \cos\alpha(y - y_0) \end{bmatrix}$$

Separating the equations,

$$x' = (s\cos\alpha)x + (s\sin\alpha)y - s(x_0\cos\alpha + y_0\sin\alpha)$$
$$y' = (-s\sin\alpha)x + (s\cos\alpha)y + s(x_0\sin\alpha - y_0\cos\alpha)$$

Substituting $a = s\cos\alpha$ and $b = -s\sin\alpha$ in the above equations, one obtains

$$x' = ax - by + c$$
$$y' = bx + ay + d$$

where the scale factor s and the rotation angle are given by

$$s = \sqrt{a^2 + b^2}$$
$$\alpha = \tan^{-1}\left(-\frac{b}{a}\right)$$

Hence one can apply the exact scale and rotation to the new map or image to register it properly to the standard image or map.

7.9 Some Well Known Map Projections

7.9.1 Mercator Projection

The most widely used Mercator projection was the first projection to be used to prepare the first atlas (a collection of maps) of the earth. Gerhardus Mercator (1512–1594) born at Rupelmond in Flanders devised this map projection in 1569 for navigation at sea. Mercator projection is a fundamental map projection falling under the cylindrical and conformal projection categories. The meridians are equally spaced whereas the parallels are unequally spaced straight lines. The scale is true along the equator.

In a map prepared by Mercator projection, the sailing route between two points at sea is represented in the map as a straight line. If the direction or azimuth of the ship remains constant with respect to north then the route is a straight line between the starting point and the destination point. These kind of sailing routes are called loxodromes—straight lines although great circle. Hence the maps prepared using Mercator projection are standard navigational tools.

By placing the X-axis of the projection on the equator and the Y-axis at longitude λ_0, the (x, y) coordinate of a point corresponding to longitude and latitude ϕ of a point on earth surface is given by the equations (7.11) and (7.12)

$$x = R(\lambda - \lambda_0) \tag{7.11}$$

$$y = R \ln \tan\left(\frac{\pi}{4} + \frac{\phi}{2}\right) \tag{7.12}$$

Equations (7.13) and (7.14) show the inverse formula for computing (θ, λ) from a given (x, y) of earth surface.

$$\phi = \frac{\pi}{2} - 2\arctan\left(e^{-\frac{y}{R}}\right) \tag{7.13}$$

$$\lambda = \frac{x}{R} + \lambda_0 \tag{7.14}$$

7.9.2 Transverse Mercator Projection

Transverse Mercator projection is a variation of the original Mercator projection where the projection surface (cylinder) is wrapped around the earth model in a transverse position i.e. the axis of the cylinder is normal to the axis of the earth.

In other words, the projection is achieved by wrapping the cylinder around a sphere or ellipsoid representing the earth so that it touches the central meridian throughout its length, instead of following the equator of the earth.

In 1772, the prolific German mathematician and cartographer Johann Heinrich Lambert inverted the transverse Mercator projection in its spherical form.

7.9.3 Universal Transverse Mercator (UTM)

Finally, the 'universal transverse Mercator projection' is a map projection, which maps the sphere representing the earth into 60 zones of 6° each, bounded by meridians of longitude (extending from the North Pole to the South Pole). Each zone is mapped by a transverse Mercator projection with the central meridian in the center of the zone—thus UTM is a global implementation of the transverse Mercator projection. Imagine an orange being comprised of 60 segments—each segment would be equivalent to a UTM zone. The zones extend from 80° S to 84° N. The UTM is depicted in Fig. 7.8. Each UTM zone, as mentioned before, is restricted to a width of six degrees of longitude. This is to avoid distortions becoming too large. The meridian at the zone's center is referred to as the central meridian. The point of intersection between the equator and the central meridian, is assigned the following values, ensuring that all coordinates within the zone are greater than zero,

East   500,000.000 meters

North   0.0   meters (Northern Hemisphere)

or

10,000,000.000 meters (Southern Hemisphere)

Under the UTM system, each east and north coordinate pair could refer to one of sixty points on earth. The same coordinate pair exists in each of the sixty zones. Consequently, the zone number must be quoted with the East and North values. The zone number effectively behaves as a third coordinate.

An example of a UTM projection is the Australian Map Grid. It lies in UTM Zones 47 to 58.

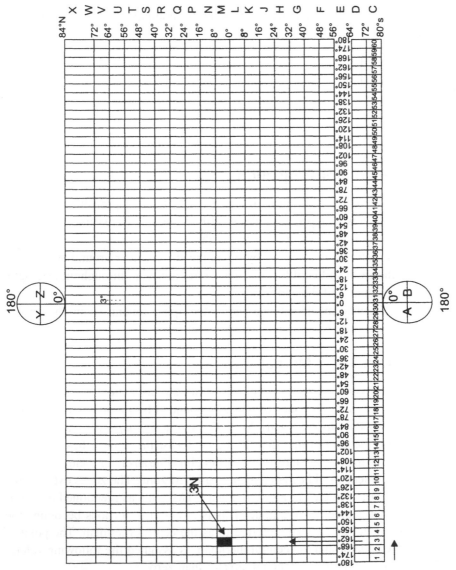

Fig. 7.8 Universal transverse mercator zones

7.10 Naming Different Map Projections

It is clear from the foregoing discussions that there exist numerous map projections depending upon the usage, the tangent surface they employ, the earth model they use and the physical characteristics they preserve. Sometimes, map projections reflect the name of the cartographer or mathematician who formulated it for first time. Hence it is necessary to categorize and have a uniform naming standard of these map projections so that the name represents all the characteristics the map projection preserves. This will make it easy for the cartographer to choose and employ a particular map projection suitable for his/her purpose.

The four major properties based on which map projections are formulated and classified are: (i) the projection surface; (ii) the intersection of the mapping surface with the model of the earth; (iii) the aspect; and (iv) the geometric property it preserves. These four properties affect the map projection predominantly and hence require careful consideration while choosing a map. To describe the name of a map projection, some property along with the name of the inventor is used. For example:

Albert's equal area conical

Lambert's conformal conic

Lambert equal area conical

By observing the above examples, a syntax for naming map projections can be formulated by suitably concatenating the aforementioned properties.

[Name of the inventor] – [Geometric property it preserves] – [Projection surface] – [Intersection of the surface]

Sometimes the most frequently used map projections are identified by the name of the inventor itself i.e.the name of the inventor is sufficient to identify the other properties of the map projection e.g. Mercator projection implies that the projection is conformal and cylindrical.

References

[1.] Bugayevskiy, M. Lev and John P. Snyder. *Map Projections: A Reference Manual. Taylor & Francis.* 1995.

[2.] Canters, Frank and Decleir Hugo. *The World in Perspective – A Directory of World Map Projections.* Chichester, England: John Wiley and Sons. 1989.

[3.] Deetz, C. H. and O. S. Adams. *Elements of Map Projection – With Applications to Map and Chart* Construction and Use of Maps and Charts. 5th ed. U.S. Coast and Geodetic Survey Spec. Pub. 68. Washington D.C.: U.S. Government Printing Office. 1945.

[4.] Kellaway, G. P. *Map Projections.* London: Methuen and Co.1946. 2nd ed. 1949; reprint 1962; 1970.

[6.] Maling, D. H. *Coordinate Systems and Map Projections.* 2nd edition. Oxford: Pergamon Press. 1992.

[6.] McDonnell, Porter W. Jr. *Introduction to Map Projections.* New York: Marcel Dekker. 1979.

[7.] Pearson, Frederick II. *Map Projection Equations*. Dahlgren, VA: Naval Surface Weapons Center, Dahlgren Laboratory. 1977. [See review by Snyder, J.P. *Amer. Cartographer* 7(2): 178–179. 1980.]

[8.] Pearson, Frederick II. *Map Projection Methods*. Blacksburg, VA: Sigma Scientific. 1984. Revision of Pearson 1977. [See review by Snyder, J.P. *Amer. Cartographer* 50(10): 1464–1465. 1984]

[9.] Pearson, Frederick II. *Map Projections*: *Theory and Applications*. Boca Raton, FL: CRC Press. 1990. Revision of Pearson 1984.

[10.] Richardus, F. *Map Projections for Geodesists, Cartographers and Geographers*. Amsterdam: North-Holland Publishing Co. 1972.

[11.] Snyder, J.P. and R.M. Voxland. *An Album of Map Projections*. USGS Professional Paper. 1453, 249 p. 1989.

[12.] Snyder, John P. and Phillip M. Voxland. *An Album of Map Projections*. U.S.G.S. Professional Paper 1453. Denver: U.S. Government Printing Office, 1989. Reprinted 1994 with corrections.

[13.] Snyder, John P. Flattening the Earth – Two Thousand Years of Map Projections. Chicago: University of Chicago Press. 1993.

[14.] Snyder, John P. Map Projections – A Working Manual. U.S.G.S. Professional Paper 1395. Washington D.C.: U.S. Government Printing Office. 1987. Reprinted 1989; 1994 with corrections.

[15.] Snyder, John P. Map Projections – A Working Manual. Washington D.C: US Government Printing Office. 1987.

[16.] Snyder, John P. Map Projections Used by the United States Geological Survey. 2nd ed. U.S.G.S. Bulletin No. 1532. Washington D.C.: U.S. Government Printing Office. 1983.

[17.] Steers, J. A. An Introduction to the Study of Map Projections. London: University of London Press. 1965. 1st ed. 1927; 15th ed. 1970.

Questions

1. What is map projection? Why is map projection required in GIS?
2. How is map projection manifested in GIS? How is it different and more advantageous than traditional cartography?
3. Give the process flow of map projection explaining the different steps involved in achieving a map.
4. What is the basis for classification of various map projections?
5. How are the different types of map projections used in obtaining different maps?
6. Derive the formulae for cylindrical map projection with the help of a figure.
7. Derive the formulae for conical map projection with the help of a figure.
8. Derive the azimuthal map projection with the help of a figure.
9. What is an affine transformation? When it is used?

8

Output Range of GIS

Sometimes similar types of information systems designed and developed by different organizations are radically different. This is because the design goals and end user application requirements driving the design are different e.g. the GIS developed for civil application is different from that designed for defense application. For that matter, the functionalities of GIS utilized for a mining application is radically different from the GIS used for navigational application. From the users' perspective, GIS from different developers differ in the way the output is generated. Hence when it comes to choosing a GIS package appropriate for one application, it is essential to compare the functionalities and outputs. Often the criteria applied for benchmarking of similar ranges of information system in general and GIS in particular, are quantity and quality of generated outputs of the system. Hence the output range plays a pivotal role in comparing and finally selecting a GIS for a specific project.

Understanding a GIS purely from its output's point of view require an exhaustive study of its output range. The outputs of an information system manifests differently depending on ways of its usage and areas of applications. For a versatile information system like GIS, which is being used in many different applications, enlisting exhaustively the outputs, is a non-trivial task. This is because GIS is nascent and is an evolving area of applied science. Besides which, unlike most information systems where the result of processed output is textual (numeric, message or a report), GIS has different types of outputs for different inputs and requirements. In this chapter an attempt has been made to enlist and categorize the output range of GIS exhaustively. To start, a formal definition of output range has been attempted.

8.1 What is Output Range of GIS?

The term GIS is fundamentally about processing, visualization and analysis of spatial data and related events in space and time. It includes the sequence of operations that are performed to process the input data, analyze, visualize, store, compute and display results in different ways. Therefore, the chronology of operations that are performed on the input data set to obtain a specific output from GIS is important. For a same set of inputs, under different circumstances, GIS generates a radically different view to space and time. It requires the ability to present multiple views of the same geographic information under different spatio-temporal conditions. Hence it can be concluded that the output of a GIS is a function of the input domain or set of inputs, the processing algorithm applied and the space and time of processing.

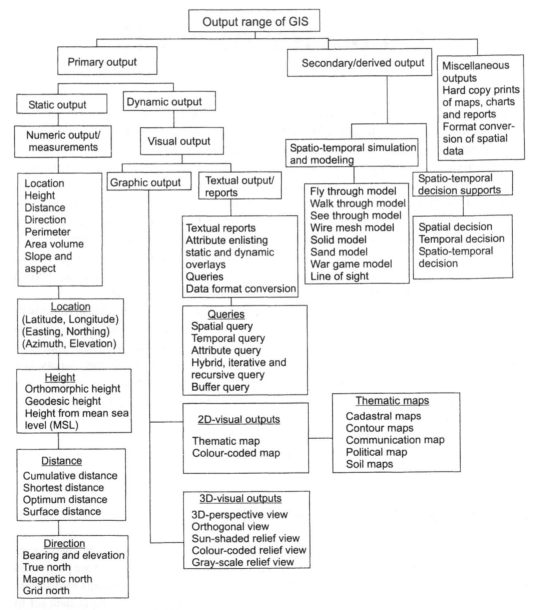

Fig. 8.1 Output range of GIS

From the foregoing discussions, one can formally define the output range of a GIS as, 'all the outputs, in different forms and manifestations, a GIS can produce after processing the spatio-temporal input domain it supports'.

The output range of a GIS can be categorized under different headings. The subset of outputs sharing common major criteria is the basis of categorization. Depending upon the content and nature of outputs, the output range of GIS can be numeric, textual, visual (graphic) etc. Dependency of the output on the time factor can make it static or dynamic, current or proposed. Mostly, the outputs of a GIS are interactive and visual in nature i.e. the visual output depends upon the spatial inputs and the action of the user at that instance of time. The outputs produced by a GIS under the static head are more or less common for all GIS software, whereas the dynamic and interactive category heavily depends upon the events generated by a specific user. Dynamic outputs manifest themselves through different processing modules and visual modules. These concepts have been explained in detail in subsequent sections.

The entire output range has been categorized under two heads (i) primary output and (ii) secondary output or derived output. An attempt has been made to put these diverse outputs of GIS under different heads and sub-heads in an organized manner as depicted in the block diagram (Fig. 8.1). The explanation and special notes regarding the categorization follows.

8.2 Primary Outputs

The primary outputs of GIS are those outputs, which are obtained by using simple geometric computation on the spatial data e.g. computation of spatial coordinates in a different frame of reference. Measurements of basic intrinsic properties of earth objects represented in the form of spatial data such as location, distance, direction etc falls under primary output. In other words, primary outputs are common derivatives of spatial data obtained after trivial geometric computation. These outputs are by and large common in all GIS products. A GIS in which such outputs are absent can be construed as incomplete.

8.2.1 Location

Computing the location of each and every point on the digital map or on a digital elevation model is very important in any GIS. Sometimes the location needs to be represented in different coordinate systems and with different precision. The accuracy of computing the location on a map depends upon the global reference frame or the global coordinate system, and the projection system applied on the spatial data.

Conversion of location (latitude, longitude) (φ, λ) expressed in (dd: mm: ss) format into (DD.sss) format is very important because the location need to be displayed in a digital display environment (the (DD.ssss) format is a more convenient form in this kind of environment). The display area in a GIS manifests itself as work space. The work space is a window of discrete coordinate matrix with each place referred to by a (row, column) pair. The value of each row and column are expressed using integers or floating points. So the (dd: mm: ss) format has to be converted to its equivalent (DD.sss) format in floating point. The conversion formula used to express the (degree: minutes: seconds) of a location in (degree decimal) form is given by

DD. sss = dd + mm/60 + ss/360

Generally the surveyed locations are expressed in the rectangular coordinate system i.e. (easting, northing) (E, N), which is a relative frame of reference. This avoids the mandatory direction suffix to latitude and longitude such as (north and south for latitude) and (east and west for longitude). An EN pair is often called a grid reference. It identifies a square kilometer or 100×100 square meter or 10×10 square meter or 1×1 square meter patch of earth surface depending upon the number of digits 4/6/8/10 it is expressed with. Refer to Fig. 8.2.

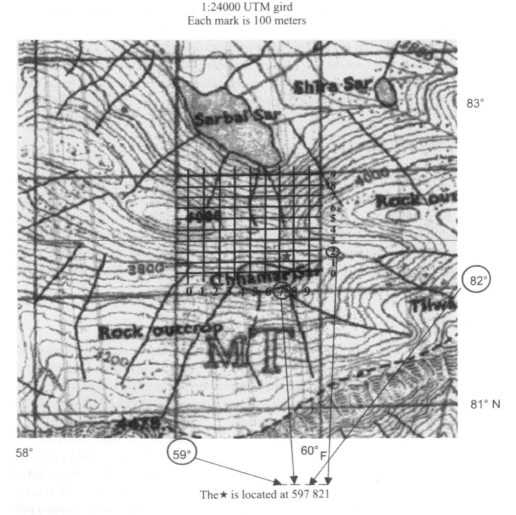

Fig. 8.2 Grid referencing with 4/6 digit easting and northing

Consider the location of the point denoted by '*'. A 4-digit grid reference using the larger square kilometer grid will give the location of the point to be 5982. If we further divide the grid to make a 100 × 100 square meter grid (sub-divisions 0, 1, 2, 3 etc marking 10, 20, 30 meter squares in each direction), we can get a more precise location of the point '*" using a 6-digit grid reference—597821. The more digits one adds to a grid reference, the more precise the reference becomes.

A convenient way of locating earth objects in polar coordinate systems is through (azimuth, distance) (θ, d). Azimuth is measured counter-clockwise from the true north of a map. The distance of the object is measured from the origin of the reference system, which is generally fixed to the geographic location of a known feature.

Computation of position of earth objects in terms of (E, N) or (x, y, z) or (latitude, longitude, altitude) or (distance and bearing) from a fixed reference position known as the origin of the reference system is primarily driven by the coordinate system used by the data generating system but a GIS has the provision of converting spatial data from one to any other system (refer to Chapter 6).

8.2.2 Height

Earth surface is not uniform and needs to be modeled so as to give a frame of reference with respect to which all geodetic measurements can be carried out. The model frame of reference is expressed in terms of vertical and horizontal geodetic datum. Horizontal datum is used for accurate positioning of earth objects relative to each other and with respect to the reference surface. The vertical datum is important for measurement of the height of objects with respect to the earth surface. The height measured depends on the vertical datum used—the height of the same object measured with respect to different model surfaces (geoid, ellipsoid or spheroid) will yield different values. To express them mathematically, different datum surfaces use different parameters. Also there is no single datum that uniformly models the whole natural earth surface. Hence there is a concept of local datum which expresses the regional earth surface more accurately, and there is a global datum (WGS 84), which averages and models the entire earth surface (For a more detailed discussion on geodetic datum, please see Section 6.7). The different datum surface, parameters expressing them and the local and global values used are enlisted in a table in Appendix C.

From the foregoing discussion it can safely be concluded that for a particular object, different datum will give different height values as output. Prominent datum surfaces used for most practical purposes are geoid, ellipsoid and spheroid. The definitions of these surfaces are given below.

Geoid

The geoid maps the equi-potential surface in the gravity field of the earth which approximates the undisturbed mean sea level extended continuously through the continents. The direction of gravity is perpendicular to the geoid surface at every point (Fig. 8.3). The geoid surface is mostly referred to for astronomic observations and geodetic levelling.

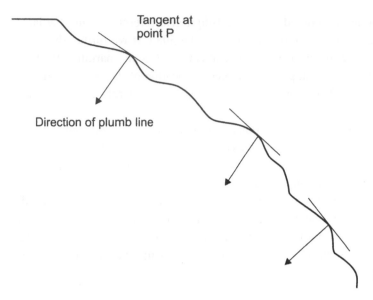

Fig. 8.3 Modeled geoid surface, perpendicular to the direction of local gravity everywhere

Ellipsoid

An ellipsoid is a mathematical figure (Fig. 8.4) generated by the revolution of an ellipse about one of its axes. The ellipsoid that approximates the geoid is an ellipse rotated about its minor axis, which generates an oblate spheroid.

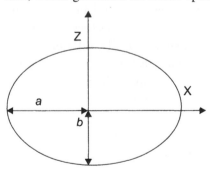

Fig. 8.4 Ellipsoid model, the shape and size determined by *a* and *b*

The height of a point above the geoid is also known as the orthometric height. Orthometric height can be positive or negative depending on whether the point is located above or below the geoid surface.

Geodesic height is measured from the surface of earth expressed through the vertical datum, where the earth surface is tangent to a plum line at every point of the surface.

All GPS use the ellipsoid as datum surface with WGS84 as the datum. Hence in case of the height measured using GPS it is computed from the reference ellipsoid, not the geoid. Therefore the height obtained from GPS and DGPS are known as ellipsoidal heights. An ellipsoidal height can be positive or negative, depending on whether the point is located above or below the surface of the reference ellipsoid. Hence ellipsoidal heights are of purely geometrical value rather than having any physical meaning. Mean sea level height (MSL) is the height measured where the earth surface is averaged as the mean sea level.

Therefore the height of a position measured with reference to the geoid and ellipsoid has two different values. The geoid–ellipsoid separation is known as the undulation of the earth surface or the geoidal height. This is depicted in Fig. 8.5. The relation between the orthomorphic height, ellipsoidal height and the geoidal height has the following relationship.

Orthomorphic height = Ellipsoidal height +/– Geoidal height

Not all GIS have the functionality to measure all three types of heights pertaining to an earth object.

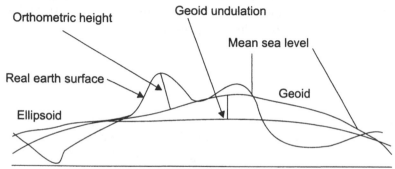

Fig. 8.5 The height due to geoid, ellipsoid and orthometric height

8.2.3 Distance

Measuring the distance between two points in a digital map in any GIS has many connotations such as cumulative distance, shortest distance, optimum distance, and surface distance etc. Not all GIS have direct functions to compute all the above types of distances. The formula and algorithms used to compute different distances are different. The cumulative distance is computed using the formula given below

$$D = \sum_{i-1,j-2}^{i-n,j-n-1} d(V_i,V_j)$$

$$d(V_i,V_j) = \sqrt{(x_i - x_j)^2 + (y_i - y_j)^2}$$

where $V_i (x_i, y_i)$ and $V_j(x_j, y_j)$ are adjacent vertices.

Shortest Distance

A digital vector map constitutes many thematic layers e.g. road, rail, power transmission line, telephone etc. All these themes can be categorized under transport network because they transport something or other from one point to another. They can also be represented in the form of a network. Road and rail network is one such instance where vehicles are the medium of transport of physical quantities. One of the most sought-after functions of a GIS is computing the shortest or optimum distance between two points in a transport network. A variant of such a function can be finding all paths connecting two points of the network or finding the optimum path or alternate path if the shortest path is rendered unusable..

The above problem is generally handled by first isolating the thematic data representing the transport network, representing the data in the form of a graph or network and finally applying an algorithm which can compute the shortest path between any two points in the network.

Many algorithms computing the shortest distance between two points in an interconnected network are available. But not all GIS have functions to directly compute the shortest and optimum distance between two points of a communication network. Generally the GIS have to be customized to carry out such functions. Many professional utilities or tools have been developed to solve such problems and are packaged into the GIS software. A generic solution is explained in this section. Generally a transport layer (road network, power distribution lines or telephone line etc.) is represented by a directed graph G (V, e),

where V = Set of nodes or vertex represented by $V (x_i, y_i)$ and

Edge $e = (v_i, v_j)$ is a set of directed arcs emanating from v_i and terminating at v_j.

In terms of data, a vertex can be a road junction, railway station or a harbour etc. representing sources and destinations. Edge e is a connected path between two vertexes such as roads, rails, air routes, power distribution lines etc. between two points on earth. The spatial data is pre-processed to remove any isolated node or point (a point which is not connected by any arc), thus preparing a directed acyclic graph (DAG) G (V, e).

The DAG is stored as a 2D-square matrix of size $n \times n$ where n = number of nodes in the DAG. This matrix is known as the adjacency matrix defined by

$a(i,j) = 0$ if V_i is not connected to V_j

$a(i,j) = 1$ if V_i is connected to V_j by an edge e_{ij}

The connected matrix is the basis for computation of all possible routes and shortest route amongst the possible routes. Well-known algorithms such as Dijkastra's algorithm and Warshall's algorithm are used to compute the shortest path. These algorithms are applied to the adjacency matrix to compute all paths and then compare and find the shortest route

Fig. 8.6 Shortest paths between start and stop points

amongst them. An example of the shortest route in a road network or communication routes is depicted in Fig. 8.6.

To find whether a path exists in a DAG G (V, e) from V_i to V_j can be answered by examining whether a path of length 1, 2, 3... exist between V_i to V_j.

This can be found by examining the expression

Path $K[i][j] = a_{[i][j]}$ II $a_{2[i][j]}$ II $a_{3[i][j]}$ ......II $ak_{[i][j]}$

where, $a_{[i][j]}$ is the adjacency matrix representing the DAG G (V, e) and $ak_{[i][j]}$ is the kth power of $a_{[i][j]}$. The operation 'II' is the logical OR operation of Boolean algebra. Hence in the above expression, the path is computed as a transitive closure of intermediate paths between the elements of adjacency matrix.

If Path $K[i][j] > 0$, then there exist a path of length k between V_i to V_j. The complete process is depicted using the block diagram in Fig. 8.7.

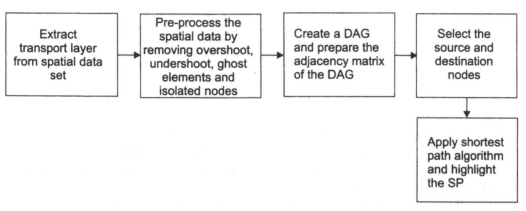

Fig. 8.7 Work flow of SP computation

The distance between two points on the earth surface representing two real objects is known as geodesic distance. The computation of geodesic distance between two points on earth crust (refer to Chapter 4) given their geographic coordinates (latitude, longitude) has many applications in navigation and guidance of aerial vehicles, weapons, ships and astronomy.

8.2.4 Direction

A sense of direction is important in air, sea or desert—this is because there are only a few landmarks in these regions, the profile is the same everywhere, and hence one can easily loose their way. The direction at any location is computed as an angle measured counter-clockwise from the true north of the map. Generally the true north is indicated as an arrow in any map. In earlier days, navigating ships in deep sea were done using the direction with respect to the pole star indicating the true north and magnetic compass indicating the magnetic north. Hence computing direction at any position was and still is the basis of navigation in sea, air or land. The direction shown by the magnetic compass is known as the magnetic north. Because of the varying magnetic field of the earth, the direction shown by the magnetic compass is different at different locations of earth surface. The direction with respect to the pole star is the true north. Generally, in a navigational map or in a navigational chart, true north is depicted diagrammatically along with magnetic north and grid north. Formally, defining the different kinds of north (Fig. 8.8) will help explain the concepts clearly.

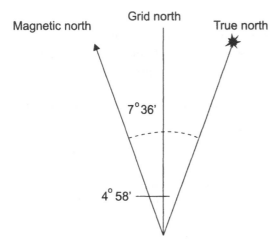

Fig. 8.8 True, grid and magnetic north

True North (**TN**): The direction of the meridian to the North Pole at any point on the map.
Magnetic North (**MN**): The direction of the magnetic North Pole as shown on a compass free from error or disturbance.
Grid North: The northern direction of the north–south grid lines on a map.
Magnetic Declination: The angle between the magnetic north and true north at any point. Sometimes, the term 'magnetic variation' is used—is mainly on nautical and aeronautical charts. Normally, however, magnetic variation is taken to refer to regular or irregular changes with time of the magnetic declination, dip or intensity.
Grid Convergence: The angle between the grid north and true north.
Grid Magnetic Angle: The angle between the grid north and magnetic north. This is the angle required for conversion of grid bearings to magnetic bearings or vice versa.
Annual Magnetic Change: The amount by which the magnetic declination changes annually because of the change in position of the magnetic north pole.

In GIS, the information regarding true, grid and magnetic north is computed in each map sheet as depicted in Fig. 8.8. The information is necessary to determine the true, grid and magnetic bearings of any line within the map sheet and is generally provided in the form of a diagram with explanatory notes. The grid bearing of a position can be computed from its magnetic bearing by using the following equation

Grid bearing = Magnetic bearing +/– Grid magnetic angle

8.2.4 Measurement of Perimeter, Area and Volume

Not many GIS have direct functions to compute the perimeter, area or volume of a piece of earth. A piece of earth is represented in the form of a closed chain of connected vertices in GIS. Thus, computing the perimeter of an earth surface needs cumulative addition of distance between two adjacent chains.

Chapter 4 explains in detail the computation of height from the datum surface; perimeter, surface area, or projected area of an object described by a closed chain of coordinates; and the computation of volume occupied by a portion of earth or a specific object. It also describes how the (azimuth, elevation, distance) of a celestial object with respect to an observer positioned at the center of the earth (geocentric earth) or on any point on the surface of the earth can be computed.

8.3 Secondary Output

Secondary outputs of GIS are also known as derived outputs. They are obtained by the specific application of a series of cartographic computations on the primary data or a combination of primary data. The outputs are in the form of visual display or textual report. Secondary outputs can be the outcome of various queries on the GIS. Sometimes the results are due to the cumulative action of an application of a spatial query and an algorithm in a cascaded manner. In a way, the results involve spatial decision support systems to address a particular problem. Hence the combination of spatial data and the order of sequence of operations are important in getting the secondary output.

Majority of the derived outputs of GIS are due to various types of queries and their results offered by a GIS. Main types of queries that a GIS can handle are spatial query, non-spatial query, buffer analysis and temporal query. Besides this, sophisticated GIS can handle many variants of these query types such as iterative query, compound query and hybrid query, fuzzy query, unstructured query etc. The power of any information system in general and GIS in particular lies in how versatile it is in allowing the user to put queries and obtain results. Hence a good GIS has intuitive ways of formulating queries. This involves an easy to use but sophisticated GUI. The GUI interacts with the user through data, voice or touch screen to intake the parameters formulating the queries.

The following sections will describe the various derived or secondary output from a GIS.

8.3.1 Spatial Query

Given the location of a feature or group of features, spatial query searches and highlights these spatial features on the map and further discovers what the feature is about. Spatial queries can be of the form:

Find and highlight all spatial feature(s) having coordinates $\{=, =, >, >=, <, <=\}$ (x, y, z);
Find and highlight all spatial feature(s) within coordinates $\{=, =, >, >=, <, <=\}$ (x_i, y_i, z_i) and (x_j, y_j, z_j).
For example, Fig. 8.9 shows a feature highlighted as a result of a spatial query.

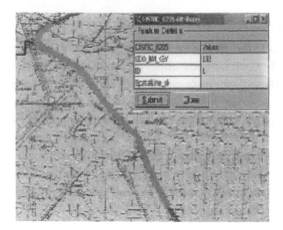

Fig. 8.9 Highlighted spatial query

8.3.2 Attribute or Non-spatial Query

Given the attribute or attributes describing the spatial feature(s), an attribute query finds and highlights where the feature is. Attribute queries can be of the form:

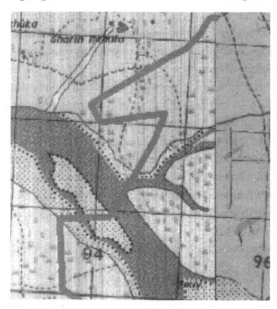

Fig. 8.10 Highlighted attribute query result

Find and highlight all spatial features having attribute a_1;
Find and highlight all spatial features having attribute a_1 {&&, II, ~} a_2;
Find and highlight all spatial features having attribute values of attribute $(a_1 = v_1)$;

Find and highlight all spatial features having attribute values of attribute $(a_1 = v_1)$ {&&, II, ~} $(a_2 = v_2)$ ...

Figure 8.10 shows the output of an attribute query.

8.3.3 Buffer Query or Buffer Analysis

Buffering is an important pre-analysis technique which is used to constrain space around individual terrain features. It combines spatial data query techniques and cartographic modeling. It is generally used to define all space within a certain distance of a type of feature, or a subset of features that are selected according to an attribute value. The user must set buffer distances.

Points, lines and polygons can be buffered; raster pixels or groups of pixels in a raster data set can also be buffered. The command for buffering and its graphic user interface (GUI) may vary from one GIS to another GIS, but conceptually, the buffer operator is a generic GIS tool.

Given the attributes or descriptions of a feature or category of spatial features, the buffer query functions to highlight and generate a buffer zone of a specific type around the feature(s) as depicted in Fig. 8.11.

Demarcate & Highlight × {km, cm, miles} {Linear, Circular, Square} around all spatial features having attribute {a} or { $(a_1 = v_1)$ {&&, II, ~} $(a_2 = v_2)$ ...}

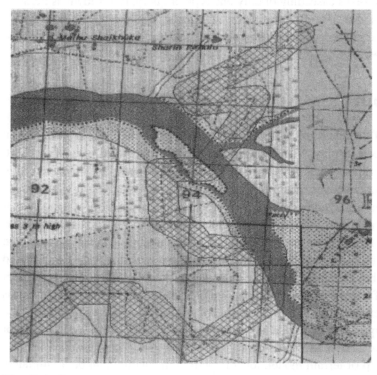

Fig. 8.11 Output of buffer query

Lines can be buffered to one side or the other as well as equal distances (right, left, and full buffers) on both sides of the line, while polygons can have an inside buffer or an outside buffer in addition to buffers on both sides of the polygon boundary.

8.3.4 Temporal Query

A temporal query is one in which the spatial features are filtered from the spatial database depending on the time or range of time of its occurrence, i.e. when they have occurred. Temporal query gives the pattern of occurrence of spatial features in a specific time or period of time. Hence it is also known as a spatio-temporal query. A temporal query can be of the form:

Find and highlight all spatial features that ccurred at <dd:mm:yy:HH:MM:SS>

Find and highlight all spatial features that occurred between <dd:mm:yy:HH:MM:SS> to <dd:mm:yy:HH:MM:SS>

A combination of the above four fundamental queries, viz spatial, attribute, temporal and buffer query can be cascaded. Cascading means the output or result sets of one query is an input to another query. It often results in a unique output. The cascading can be done for a similar query or for different types of queries, e.g. the result set of a spatial query is input to non-spatial query, whose result set can be input to a buffer query to highlight and display the results on the map. If it involves cascading of only one type of query, it is known as iterative query. If the cascading is done for different types of queries, it is called a hybrid query. Finally the result set of the entire hybrid or iterative query can be filtered for a specific time period or time of occurrence.

8.4 Visual Output

Effective interpretation of the results of a GIS analysis by a user depends to a large extent on the design of the on-screen display. Visualizing information in the form of a map, chart, image, or other form of graphic takes advantage of humans' most perceptive sense—vision (Batty 1994). Visual display uncovers patterns, associations, and processes that would likely be missed if the data were displayed in tabular or some other non-graphic form. However, it is this very ability that might lead a viewer of a graphic display to be as easily misled as informed by the contents of the graphic. A slight alteration of the colour scheme or classification method of a map, for example, might produce two different graphics that can be interpreted differently, even though they represent the exact same data. Thus it is crucial that GIS users be aware of the power of a graphic image to communicate ideas and of the considerations necessary to design an effective on-screen visual display. Cartographers have developed elegant modeling techniques that are used as guidelines by different GIS developers for unambiguous cartographic display of spatial data.

By just observing a map or terrain model, lots of spatial information can be assimilated. That is why, GIS is gaining popularity as a powerful visual information processing system. Besides the numeric results and textual tables, a sizable range of the output is produced in

GIS visually. The visual output can be categorized under graphic outputs and textual outputs. The graphic outputs can be categorized under two major heads namely (i) 2D display and (ii) 3D display. A sample set of such outputs is given below.

8.4.1 2D Display

Majority of the outputs in GIS are in the form of 2D display. The most important of these are generic processes resulting in 2D display e.g. thematic composition of 2D digital maps, generation of relief maps, colour coded maps etc. To interpret these outputs, a supplementary data is generated such as a range table or legend. A range table or legend depicts the association of sub-range of spatial data with numeric value or colour code. This helps in interpreting the 2D display. The generic process of generating the 2D display of spatial data is given in Fig. 8.12.

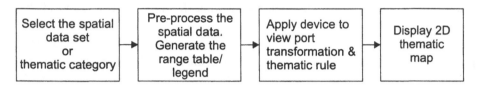

Fig. 8.12 Work flow of a 2D display

Some of the prominent outputs generated in GIS following the above sequence of operations are sun-shaded relief; colour-coded relief, cadastral map, land use map etc.

8.4.2 3D Display

The most natural and realistic view of terrain surface is through a 3D display of digital spatial data. In 3D visualization, the user gets a sense of height and undulation of the terrain surface in addition to the topology of ground features. All the analysis that are being carried out in 2D display can be carried out in 3D display also. 3D display makes use of DEM/DTED data in addition to the vector and raster spatial data usually exploited by GIS systems. The generic process of 3D visual output is depicted in Fig. 8.13.

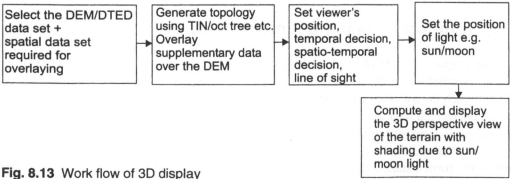

Fig. 8.13 Work flow of 3D display

The outputs that are generated using the above generic sequence of operations in GIS results in a number of important views and simulations such as the orthogonal view of terrain which gives the view of the terrain as if seen from deep space and the sun-shaded relief map depicting the pattern of undulation through shades due to obstruction of sunlight by relief of the terrain. By assigning a range of colour to different ranges of height, the generic pattern of height is generated. This generic pattern is known as the colour-coded relief view of a terrain surface. The user can have a 3D perspective of the terrain with different lighting conditions and overlay to correlate height, slope and aspect of the terrain surface.

A continuous updated view of the undulated terrain from an aircraft as seen by the pilot (fly through simulation) can be generated using the above data set. Similarly, the terrain profile, line of sight between the observer and object can be computed and visualized in 3D view. The spatial resolution of DEM and DTED obtained are quite coarse and not sufficient for simulation of a fly through. For simulation of a walk through, the resolution of data needed will be of one meter by one meter. The day is not far where, with the advent of space imaging and satellite technology, the user will be able to continuously monitor and obtain high resolution DEM data to simulate a walk through in an alien land. Further, the GIS will be capable of superimposing the environment data to have a virtual see through of the 3D terrain model.

Graphical depiction of natural phenomena occurring over the earth surface can be best visualized by depicting the phenomena over a 2D or 3D digital map as the background. Phenomena such as temperature, pressure, wind speed, direction of ocean current etc. can be best depicted with a digital map in the background. Phenomena such as plate tectonics (movement of continental plates or continental drift), volcanic eruption; seismic hazard zone can be best studied when simulated using DEM data.

Charts and graphs generally depict the numeric occurrence of any event with respect to time and space. These charts and graphs can be depicted textually as a chart or report. Graphs and charts give a statistical measure of an event or feature with respect to space and time e.g. census report, occurrence of rare earth materials in the coastal zone of India since 1960 to 2006. GIS can generate both graphical charts and equivalent textual reports with thematic overlay in 2D and 3D digital map displays.

8.5 Temporal Classification of Output

All the outputs of a GIS can be broadly categorized into two categories depending upon the mode in which the result set behaves in time, viz. static and dynamic. Static outputs are those which do not change irrespective of the current inputs and the user events. They are fairly unchanged over a longer period of time, whereas dynamic outputs change due to the user input, data and events to be visualized by superimposing the spatial data. Perfect examples of static data are the hard copy maps produced by survey organizations. The associated tables too form a static data, e.g. the population index table and the mineral resource table.

Examples of dynamic outputs are the 3D perspective of the undulated terrain depending upon the viewer's position. Enlisted below are some static and dynamic outputs produced by GIS.

8.5.1 Static Outputs

Hard copy maps, analysed tables enlisting both spatial and non-spatial data e.g. population index tables, which gives a correlation between the country, its population, growth rate with infant mortality and education are all examples of static outputs. Such table aids a planner to focus resources for the betterment of the phenomena under consideration. Some of the initial study wherein GIS has been used to understand the spatial distributions of infant mortality, literacy and poverty are being studied and analysed (Andes et al. 1995).

Infant mortality and literacy report

Nation	Population in lakhs	Rate of growth	Life expectancy	Infant mortality rate	Child malnutrition	Primary school	Illiteracy rate enrolment

Poverty study report

Name of the country	Per capita income in thousands	Child malnutrition	Infant mortality rate	Primary school enrolment rate

Generally these tables are the textual report which can be depicted as thematic overlays over a spatial map. GIS can find the national and regional average index of a property (represented as a field in the table) and generate a thematic map with colour codes giving a uniform colour to a range of average values. A legend is generated to assist the user to understand and interpret the data spatially along with the hard copy map. From the above tables, the civil administrators can establish a correlation between poverty and infant mortality rate. The area, which requires urgent attention to improve the situation, can be visualized in the map and resources can be mobilized.

8.5.2 Dynamic Outputs

The dynamic outputs of a GIS are the outcome of the processed spatial data or combination of spatial data effected by the current user input and event. The output is categorized as dynamic because most of its quantity, quality and visualization depend upon the current user input and event. Sometimes the dynamic output range gives a simulated output of the spatio-temporal phenomena. Spatio-temporal simulations are emerging functions and are being used as an aid to decision makers in many respects. Given in the following sub-sections are some typical spatio-temporal simulations.

Spatio-Temporal Simulation

Simulation of natural and man-made phenomena using spatial or time as control parameters of the simulation is known as spatio-temporal simulation. Simulations can be categorized under spatial simulation, temporal simulation and spatio-temporal simulation depending upon the governing parameter which are only spatial, only temporal or both respectively. The most popular simulations under these categories are wire frame model of terrain, sand model of the terrain, walk through model of terrain, fly through simulation etc. The widely used spatio-temporal phenomena in GIS are simulations of combat. Many war games and combat models have been simulated using GIS (Skidmore 2001). This gives the battle commander access to the battle zone, deploy his forces vis-à-vis the opposing forces and carry out a real-time simulation of the battle scenario to study the attrition of the force levels.

Of late GIS offers the best platform to simulate disaster due to weapons of mass destruction (WMD) such as nuclear, biological and chemical warfare. With GIS, numerous simulations have been carried out to see the effect of a nuclear explosion in an area and how the emergency response can be planned. Simulations due to natural disasters and planned disasters such as flooding, bombing etc are best studied using GIS. Given a point of breach on a water body and the rate of flow of water, a GIS can simulate flooding patterns by generating the area to be flooded and progress of flooding pattern.

Given a point on a terrain, the almanac data pertaining to the region (time of sun set, sun rise, moon set and moon rise) can be computed using the relative position of the celestial body with the earth. Also at a particular point of time, GIS can compute how the terrain looks like under the illumination condition due to sun, moon etc. Computing the azimuth and elevation of a celestial body from a given point on the terrain surface finds many applications in astronomy. A range of outputs that are generated can be categorized under spatial decision support. These fundamentally depict the dynamic situation of a phenomena depending upon space and time and hence can be categorized as dynamic outputs too.

Spatio-Temporal Decision Support Systems (SDSS)

The decisions that can be derived from a GIS can be categorized into (i) spatial decision (ii) temporal decision and (iii) spatio-temporal decision. These categories of outputs are considered as secondary or derived outputs of a GIS system (Birkin et al. 1996).

The secondary outputs of a GIS are the outcome of application of algorithms and a series of business logic on primary output or a set of primary outputs. The best examples of secondary outputs besides the outcome due to queries are the spatial decision support systems (SDSS). SDSS will be explained in Chapter 10 of this book.

Searching for a suitable site on the terrain, given a set of criteria is an important terrain analyses being carried out in GIS (Goodchild 1987). Finding a suitable site for locating an air defence (AD) gun in the field or fielding a network of sensors or RADARs require a careful study of the terrain under consideration. A commander often comes across such problems when deployment has to be carried out in an unknown terrain. Some of the frequently occurring decisions he would have to make are:

Find a suitable site for staging of forces while on the move.

Find a suitable site for staging of guns, sensors, and RADARS etc.

Find a suitable point of breach for effective flooding to be carried out over an area.

Find a suitable site for construction of helipads in an unknown terrain.

For a civic body like a town planner or city planner, finding suitable sites for locating civic facilities e.g. fire stations, hospitals, new settlements, airport etc. require a careful study of the terrain surface. Lots of parameters are taken into consideration to find the most suitable location so that the maximum population can optimally utilize the facility in an easy manner.

The simulated outputs of a GIS are dynamic because it varies for different spatio-temporal data and conditions imposed by the man-made structure. The use of GIS is increasing among the defence and civil communities because of the spatio-temporal analysis that a GIS can carry out.

8.6 Miscellaneous Outputs

Some outputs of GIS cannot be categorized into any of the above groups and hence are enlisted under miscellaneous category. Nevertheless these are essential outputs without which a GIS is unacceptable and commercially not viable to the user community. Hard copy outputs like prints, charts etc. falls into this category. These are essential because they have to be carried to the field by the user as references, where a sophisticated GIS is not available. Also it is impractical to take a GIS system everywhere during field operations. Hence these hard copy outputs play a significant role as a ready reference to the user. Beside this they are treated as reserve and used at the time of non-availability of GIS.

8.6.1 Overlays

One of the important segments of GIS outputs is user-created overlays. Overlays are basically scenarios of a particular phenomena occurring on a terrain. Overlays can be tactical or terrain change reports depicted graphically (Fig. 8.14). Depending upon the temporal behaviour of the phenomena, the overlays can be categorized as static or dynamic. Typical examples of an overlay are deployment of battle elements, location of corporate offices, location of civic facilities such as fire stations, hospital etc. over a geographic zone. Overlays give the spatial pattern and occurrence of the objects with map in the background. The following fall under overlay category.

1. Maps and overlays, thematic maps
2. Query outputs as thematic layers.
3. Textual reports, navigation charts, graphs
4. Collaborative view of management information system (MIS) data with map as the background

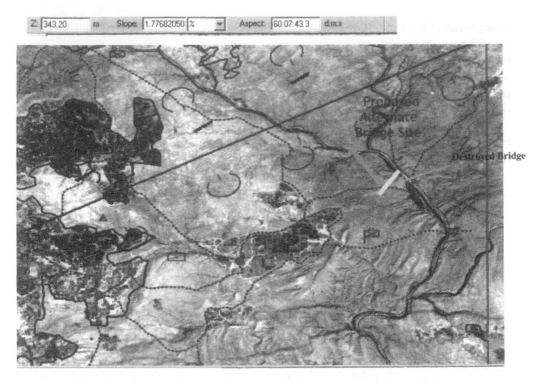

Fig. 8.14 Deployment of forces as an overlay; Computed values of slope, aspect and height from MSL at the cursor location (Also see Plate 4)

8.6.2 Spatial Data Conversion

With so many diverse devices engaged in capturing spatial data, even the format these data are stored in, are different. Some scanners store the output images in TIFF (tagged information file format), some in GIF (geographic image format) etc. The organization of the data produced by a device is known as the data format. There are lots of data formats prevailing in the digital world. The output of one device has to be consumed by another such as the scanned output may be digitised using a digitising software such as Microstation or CAD CORE. The output of Microstation is in DGN (design file) format. DGN data need to be displayed in a computer monitor using GIS software. One can safely conclude that to understand and display data produced, is an art in itself. Often data formats are proprietary to the user organization. Different data formats are practiced mainly because of the following reasons:

To protect data from unauthorized user/device

To control the use of data

But now the economics of data sharing has driven various organizations to come out with standard data formats. The simple reason is that the data produced by a particular organization should be used and understood by the intended organization, device or software. There are two common data formats used presently—raster data and vector data.

Raster data is stored in a matrix of pixels with a header defining the number of rows, columns and pixels per line etc as meta data regarding the organization of the file.

Vector data are discrete coordinates organized as points, lines and polygons with delimiters in between features. The colour, line style, width information along with geo-coding information is also incorporated in each feature. If the vector map data is attached with attributes then there is a unique number corresponding to the attribute table which is attached while digitisation of the feature. This unique number is the key to identify the attribute table.

With raster data having meta-files and vector data having unique identifying numbers, it becomes necessary to decode and decipher the data format of one system to be used by another software or system. Hence data format conversion and importing and exporting of data from one format to another is an important requirement of GIS. The data format understood and produced by a GIS is known as its native data format.

There can be four distinct types of data conversion process:
1. Vector to vector (DVD 2 DXF), (DGN 2 DVD)
2. Vector to raster (DVD 2 GIF), (DVD 2 TIFF)
3. Raster to vector (TIFF 2 DGN), (JPEG 2 DXF)
4. Raster to raster (TIFF 2 GIF), (TIFF to JPEG)

8.6.3 Hard Copy Output

The third miscellaneous category of outputs of a GIS is the hard copy of maps, DEMS or charts obtained through a printer or plotter. All major commercial GIS systems have these facilities; they only differ in the look and feel or standard operating procedure (SOP) they employ in generating the hard copy output. Following are some of the common hard copy outputs generated by GIS:
1. Printing and plotting of digital vector map.
2. Printing and plotting of thematic maps or overlays for specific operation.
3. Plotting of sea route, air route or land way points.
4. Plotting of navigation charts, night charts
5. Printing of textual reports with respect to an thematic overlay

References

[1.] Abel, D. J., P.J. Kilby, and J.R. Davis. The system integration problem. International Journal of Geographical Information Systems 8:1–2. 1994.

[2.] Andes, N. and J. E. Davis. Linking public health data using geographical information system techniques: Alaskan community characteristics and infant mortality. Statistics in Medicine 14:481–90. 1995.

[3.] Anselin, L. and A. Getis. Spatial statistical analysis and geographic information systems. In Geographic Information Systems, Spatial Modeling, and Policy Evaluation edited by M. M. Fischer and P. Nijkamp. (eds). 35–49. Berlin: Springer. 1993.

[4.] Batty, M. Using GIS for visual simulation modeling. GIS World 7(10): 46–8. 1994.

[5.] Birkin, M., M. Clarke, G. Clarke and A.G. Wilson. Intelligent GIS: Location Decisions and Strategic Planning. Cambridge (UK): GeoInformation International. 1996

[6.] Brimicombe, Allan. GIS, Environmental Modeling and Engineering. George Green Library. 2003.

[7.] Densham, P. J. Spatial decision support systems. In Geographical Information Systems: Principles & Applications edited by D.J. Maguire, M.F. Goodchield and D.W. Rhind. 1: 403–12. Harlow: Longman / New York: John Wiley & Sons Inc. 1991.

[8.] Federa, K. and M. Kubat. Hybrid geographical information systems. EARSel Advances in Remote Sensing 1:89–100. 1992.

[9.] Goodchild, M. F. A spatial analytical perspective on geographical information systems. International Journal of Geographical Information Systems 1:321–34. 1987.

[10.] Goodchild, M. F., R.P. Haining and S. Wise. Integrating GIS and spatial data analysis: problems and possibilities. International Journal of Geographical Information Systems 6:407–23. 1992.

[11.] Maguire, D. J. Implementing spatial analysis and GIS applications for business and service planning. In GIS for Business and Service Planning edited by P.A. Longley and G. Clarke. 171–91. Cambridge (UK): GeoInformation International: 1995.

[12.] Skidmore, A. Environmental Modeling with GIS and Remote Sensing. Taylor & Francis. Hallward Library. 2001.

[13.] Steyaert, L. T. and M.F. Goodchild. Integrating geographic information systems and environmental simulation models. In Environmental Information Management and Analysis: Ecosystem to Global Scales edited by W.K. Michener, J.W. Brunt and S.G. Stafford. 333–57. London: Tayler and Francis. 1994.

[14.] Warren, I. R. and H.K. Bach. MIKE 21: A modeling system for estuaries, coastal water and seas. Environmental Software 7:229–40. 1992.

Questions

1. What is the output range of an information system?
2. What is the output of a GIS? Explain with generic categories and sub-categories.
3. What is the primary output of a GIS?
4. What is the secondary or derived output of a GIS? Explain with examples.
5. Give the process flow of computing the shortest path between two points of a communication network. Explain its various steps.
6. What are the different types of queries a GIS can intake and process?
7. What is temporal query? Explain with an example.
8. What is the visual output of a GIS and its different types?
9. Give the block diagram of a work flow involving 2D and 3D display in GIS.
10. Give the temporal classifications of GIS outputs.
11. What are spatio-temporal simulations?
12. What is spatial DSS? What are its building blocks?

9

Computational Geometry Used in GIS

A technique totally different from mainstream subjects has the potential to significantly influence and modify the mainstream discipline itself. In the present context, the recent developments in computational geometry (CG) have the potential to change many of the ways we look at problems involving spatial processing in GIS. CG has a profound impact in the discipline of digital cartography, geomatics, or geographic information systems (GIS). In fact, a nascent area of research is emerging known as computational geography.

Terrain data is highly unstructured. Data structures used in GIS such as point, line, triangle, polygon etc. have inherent problems and limitations while representing and processing terrain data. In such cases, data structures such as convex hull, Delaunay triangulation, Voronoi diagram, quad tree, oct tree etc. provide an attractive alternative. This is because these data structures are continuous and space-filling structures, eliminating many problems of traditional data structures in spatial data modeling. They also adapt easily to different forms in different dimensions thus overcoming the handicap of traditional GIS operations.

In this chapter, the role of computational geometric algorithms (CGAL) has been explored (Mark et al.). To start with a formal definition of CGAL with an explanatory relevance to GIS is enunciated. Followed by the definition, the contents of CAGL, the spatial data structure and algorithms acting on them have been discussed. A comparison of spatial data structure with geometric data structure has been carried out to clearly bring out the role of spatial data structure in GIS. Important candidates of SDS and CAGL have been further elaborated with respective algorithms for the benefit of the readers. A variety of GIS problems are described, along with possible solutions using computational geometric algorithms (CGAL). CGAL is used a lot in building and analysis of queries involving spatial data. This aspect of application of CGAL in spatial query is discussed in a separate section. The real challenge for the future is not only to implement more efficient versions of traditional GIS functions, but also to see how the two disciplines can combine to produce a completely new functionality.

9.1 Definition of CGAL

The first principle of GIS is to model, view, analyze and control (MVAC) spatial objects. This is different from the traditional modeling paradigms employed in the design of general information systems, where model, view and control (MVC) are mainstay practice. In GIS, earth objects are modeled geometrically and stored as spatial data using traditional geometric data structures such as points, lines, triangles and polygons etc. These are in turn visualized using cartographic display rules before analysis. The user control over the data is always through display and query.

Computational geometry is a set of data structure and algorithms designed for modeling and analysis of problems involving geometric objects (Preparata et al.). Since it helps in modeling and analysis, CG algorithms play a pivotal role in every stage of MVAC of spatial data. Applications of CGAL is evolving and increasing day by day. The block diagram in Fig. 9.1 provides a glimpse of the usage of CG in diverse areas of application. Thus it can be seen that computational geometric algorithms are capable of addressing many problems arising in application areas such as computer graphics, computer vision, human–computer interface, astronomy, spatial information processing and geographic information system.

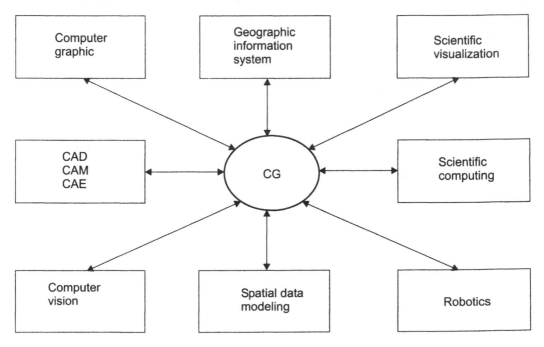

Fig. 9.1 Collaboration of CG with different areas of application

9.2 Role of Computational Geometry in GIS

Geographic information systems offers functionality to model, store and manipulate spatial data. In this, it brings together two different branches of computer science i.e. (1) database technology which offers methods and structures to store, retrieve and manage spatial data and (2) computational geometry, which offers algorithms to efficiently analyze and manipulate spatial objects. One of the most important developments of computational geometry is its application in GIS. CGAL is capable of modeling and analyzing the most complicated spatial data efficiently using various spatial data structures and algorithms. In a way, the spatial processing engine of GIS can be thought of as a library of computational geometry algorithms (CGAL). The CGAL manifests itself as spatial data structures holding terrain objects and algorithms applied over these data structures. They can hold, query, retrieve, analyze and visualize spatial data in an efficient manner in different stages of MVAC.

9.3 Contents of CGAL Library

From the foregoing discussions it can be safely concluded that the CGAL library constitutes two parts viz.

1. Spatial data structure which hold spatial data describing the terrain objects as well as geometric data structures representing geometric features.
2. The computational geometric algorithms that are applied on the spatial as well as geometric data structures for computation, analysis and visualization.

Further, the algorithms can be categorized into three categories viz. (i) geometric primitives, (ii) geometric predicates and (iii) geometric algorithms (Frank 1992). Primitives are result-computing functions; predicates are decision-generating functions applied over the geometric data; and geometric algorithms hold and analyze spatial data for query and analysis over a wider geographical range. The constituents of CGAL and their categorization are depicted in Fig. 9.2.

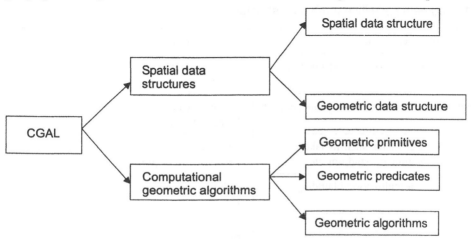

Fig. 9.2 Categorization of CGAL into DS and algorithms

The spatial data structure models and holds all data which has coordinate data associated with them. Geometric data structures are those which holds structured geometric data such as point, line, poly line, polygon, polyhedron etc. Thus while geometric data structures represent and hold data which are structured, spatial data structures can represent and hold data which are unstructured e.g. convex polygon, tessellated surfaces such as Delaunay triangulation, Voronoi tessellation, quad tree, oct tree etc (Bowyer 1981)—in other words, geometric data describing an object are highly structured whereas the spatial data describing terrain objects are random and unstructured. The main difference between a geometric data and spatial data is that the coordinates of spatial data are associated to a position on earth surface through latitude, longitude and height whereas the geometric data has coordinates arbitrarily enumerated according to the device generating them.

Algorithms, which generally act on geometric data structures, deriving quantitative outputs such as position, distance, perimeter, area, volume etc are called primitives. These are functions, which generate single output and take the geometric data as the input through its arguments. Hence, a geometric primitive generates a numeric output pertaining to the geometric object. Primitives are generic and can be applied both for 2-dimension or 3-dimensional spatial data.

A geometric predicate is a function which intakes one or more geometric objects as argument and returns a decision in the form of a Boolean (true/false) output. Hence predicates enforce semantics to the spatial data (between one or more geometric objects), whereas geometrical primitives generate a quantitative output of a geometric object.

A spatial data structure imparts ordering among spatial objects thereby establishing a topology. A computational geometric algorithm operates on a mixture of geometric and spatial data structures and generates a spatial ordering among the spatial objects. Further, the spatial object or objects under consideration can be visualized, analyzed and stored using CGAL (Carlos Morino 1998). Enlisted in Tables 9.1–9.2 are a candidate list of geometric primitives, predicates and algorithms frequently used in GIS.

Table 9.1 Geometric primitives with data structures

Geometric representation	Geometric primitives
Point = Position of a house, road junction	Position (point) Height (point)
Line, poly line = Communication elements such as road, rail	Distance (line) Direction (line)
Polygon = Area elements such as lake, administrative boundary, stadium etc.	Area (Polygon) Perimeter (Polygon)
Volex = A tower, a multi-storied building	Volume (volex) Height (volex)
Triangle	Area (triangle) Perimeter (triangle)

Table 9.2 Geometric predicates with decisions

Geometric decision	CG predicates
Whether two lines L_1 and L_2 intersect properly or not?	Line–line intersection Intersect (L_1, L_2)
Whether three points are collinear or not?	Point collinear (p_1, p_2, p_3)
Whether the points p_1, p_2 and p_3 are oriented clockwise or counter-clockwise (CCW)	Orientation: CCW (p_1, p_2, p_3) or CW (p_1, p_2, p_3)
Whether the point p lies inside line L, triangle T, polygon Poly, circle C	Inside (p, L) Inside (p, T) Inside (p, C) Inside (p, Poly)
Whether the coordinates of a triangle form a counter-clockwise orientation	CCW (T) CW (T)

A careful study of the primitives and predicates in the tables brings out two important conclusions—that the CGAL has spatial data structure to represent basic geometric primitives such as points, lines, poly lines etc. and geometric operations such as spatial intersection, touching, nearest neighbourhood, containment etc; and the predicates help in deriving important decisions regarding the spatial objects, giving the relative occurrence of these objects in space.

9.4 Data structures used in GIS

Points, lines, triangles and polygons are basic geometric data structures which impart order and topology to terrain features modeled geometrically. Over the years data structures have been designed which describe self-adjusting random patterns in two and three dimensions. Tree is one such data structure which expands itself to hold data which are spanning recursively. There are many variant of trees. Out of them, binary tree, red–black tree (RB tree), quad tree, oct tree are special data structures which are extensively used in GIS for holding spatial data. In the following sections, triangle and quad tree data structures are taken as candidate examples to explain the concept of spatial data structures. Also discussed are algorithms which will load spatial data to the respective data structure.

Table 9.3 lists some important data structures and computational geometric algorithms that are frequently used in GIS. This table is only a sample and not an exhaustive list of the problems encountered in GIS.

Table 9.3 CGAL with relevant GIS functions

Sl No	Problems in GIS	Data structure
1	Searching the location of an object point inside a polygon	Triangulation of planar domain TIN (triangular irregular network), Triangle data structure Voronoi polygon or Thessian polygon
2	Finding a particular bridge on a road or river	Point on a line
3	Finding the intersection of a road with river	Line–line intersection
4	Finding the flood affected zone	Polygon inside a polygon
5	Find the maximum area affected by flood	Convex hull computation
6	Find the amount of water contained in a reservoir	Convex volume representation
7	To find the shortest/optimum/ safest/ optimum path from one point to another	Directed acyclic graph (DAG) Connected matrix Adjacency matrix Weighted adjacency matrix
8	To find the safest path and cross-country route in an area	Polygon
9	Multi-resolution representation of planar map	Quad tree representation
10	Multi-resolution representation of 3D terrain or 3D space	Oct tree representation

9.4.1 Triangle DS

A convenient data structure to bind and hold three vertices, edges and neighbour triangles to a single unit is triangle data structure. It is frequently used in representation of terrain surface using TIN (triangular irregular network)—a network of triangles. A schematic diagram of a triangle data structure is depicted in Fig. 9.3. It is also defined in C language as:

```
Typedef struct pt_3D
{
float x;
float y,
float z,
} Point; // data structure defining point in 3D
Point nodes[MAX_NODES] //Array MAX number of points

Typedef line{
Point p1, p2;
Float length;
} Side;
Side S[MAX_NODES/2] // Array of sides of type line segment
```

Typedef tri {
Point p_1, p_2, p_3;
Float area, perimeter;
}Triangle; // data structure defining point in 3D
Triangle T[MAX_NODES] // Array of MAX number of triangles

Typedef triD {
    Triangle T_i; // index to the triangle
    Point p_a, p_b, p_c; // index to vertices of triangle T_i
    Side s_a, s_b, s_c; // index to the respective sides
    Triangle T_a, T_b, T_c; //index to adjacent triangles
    } triangleDS;

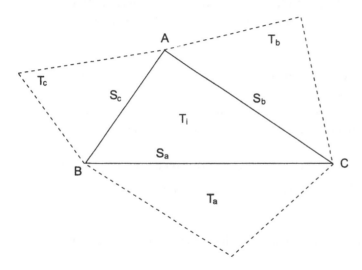

Fig. 9.3 Triangle data structure

9.4.2 Quad Tree DS

2D polygonal objects, which occur frequently in a digital vector map, can be better represented using quad tree data structures. A quad tree achieves coherent representation of 2D maps by sub-division of non-homogeneous objects to smaller squares until a minimum scale is reached or homogeneity obtained.

In case of quad tree, the internal representation of a terrain object is a tree. Quad tree avoids the necessity of large amount of storage space by linear coding of leaves recursively. These data structures also bring to the video memory only that portion of the large data, which is in the immediate view field. Hence it is used mostly in game design, fly through and walk through simulations.

A quad tree is an adaptive tree structure, which sub-divides the 2D space containing the data into four identical quadrants (as its nodes) recursively until all nodes either contain only a single object, are empty, or a minimum scale of sub-division is reached (Fig. 9.4(a)).

From the definition it is clear that each node of a quad tree has four successor nodes. The intermediate nodes are space containing compound elements or more than one element. The root node is the overall collection of objects or the container. The terminal nodes (leaves) are either empty nodes or nodes containing a single object or the node at which sub-division has reached a desired minimum scale.

In GIS, the container or root node of a quad tree is either a digital vector map containing all features modeled as points, lines, polygons or a digital image consisting of an array of pixels.

The nodes of a quad tree at each level can be represented as a record. If the node is not a leaf of the quad tree, it contains pointers to four successive nodes indexed as NW, NE, SW and SE as shown in the 2D-representation (Fig. 9.4(b)). The fields of the record are the bottom left coordinates (x, y) of the quadrant and the quantitative representation of the features contained inside the quadrant. Depending upon the contents, the nodes can be categorized into (i) white (W) node where there is no object, i.e. it is a blank node, (ii) gray (G) node which contains more than one object and which can be further sub-divided, (iii) terminal (T) node which may contain one or more objects but has reached the minimum scale of division, (iv) black (X) node, which contains only one object and needs to be further sub-divided.

With these definitions of nodes the recursive tree construction algorithm is given by:

PROCEDURE BUILD_QUAD_TREE *(x, y, min_scale, list of objects)*

For i = 1 to 4 do

List(i) = determine the objects (points, lines, polygons) that describe the heterogeneous object.

Temp_scale = scale / 2

Case (no_of_elements_of (list (i)))

0 : putout(W) // no elements → white terminal node
1 : if (scale > mscale) putout(X) // Only one element and scale is large→ Black non-terminal node

Build_Quad_Tree($x + a(i) \times$ temp_scale, $y + b(i) \times$ temp_scale, min_scale, list of objects of list(i))

>1: putout(G) // Gray node

if (temp_scale > min_scale)

Build_Quad_Tree($x + a(i) \times a(i) \times$ temp_scale, $y + b(i) \times$ temp_scale, min_scale, list of objects of list(i))

End if

End case

END PROCEDURE

Where $a(i) = \{0, 1, 1, 0\}$ and $b(i) = \{1, 1, 0, 0\}$ are arrays.

A quad tree has the following properties

- Each node may have up to four daughter nodes, corresponding to the four sub-squares that are obtained by dividing in half along each of the Cartesian direction on which the tree is defined. The sub-squares decompose space into adaptable cells.
- Each cell (or bucket) has a maximum capacity. When the maximum capacity is reached, the bucket splits.
- The tree directory follows the spatial decomposition of the quad tree.
- Each node has pointers to the daughter nodes indexed: quad [NW], quad [NE], quad[SW], and quad[SE].
- The point is of type DataPoint, which in turn contains:
  - Name
  - Coordinates (x, y)

Oct tree is an adaptive tree data structure which has an algorithm that treats each body independently. The algorithm begins by partitioning space into an oct tree, that is, a tree whose nodes correspond to cubical regions of space. Each node may have up to eight daughter nodes, corresponding to the eight sub-cubes that are obtained by dividing in half along each Cartesian direction. The tree is defined by the following properties:

1. All terminal nodes of the tree have either one or zero bodies.
2. All nodes with one or zero bodies are terminal.

The oct tree provides a convenient data structure which allows us to record the properties of the matter distribution in three dimension. It is especially convenient for astrophysical systems because it is adaptive. That is, the depth of the tree adjusts itself automatically to the local object density. Oct trees are somewhat difficult to represent two-dimensionally i.e. draw on paper. The corresponding analog in two dimensions is a quad tree represented in Fig. 9.4(a) and (b).

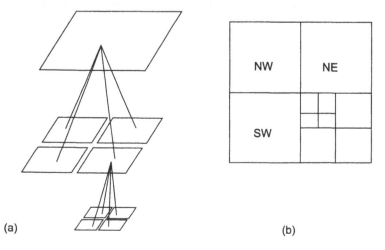

(a)   (b)

Fig. 9.4 (a) Inverted quad tree; **(b)** 2D representation of a quad tree

9.5 CG Algorithms used in GIS

9.5.1 CG Primitives and Predicates for Simple Computations

The algorithmic steps of some important computational geometric primitives and predicates often used in computation and processing of spatial information are given in this section (Mark et al.). They are expressed both in the form of Boolean expressions and mathematical predicates. The proofs of these predicates are fairly simple.

Predicate to find if three points are collinear or non-collinear

Consider the given points A (x_1, y_1), B (x_2, y_2) and C (x_3, y_3) in a plane. They are collinear if the area subtended by ABC is zero. If the area is positive then they are oriented counter-clockwise (CCW) else oriented clockwise (CW). Hence the primitive of the sign of the area itself tests the collinearity as well as the orientation of the three points. Moreover the predicate need only compare the sign of the area; not the magnitude.

Twice Area (A, B, C) = $((x_1y_2 - x_2y_1) + (x_2y_3 - x_3y_2) + (x_3y_1 - x_1y_3)) = \Delta$

    If $\Delta = 0$ then A, B and C are collinear

    If $\Delta > 0$ then CCW i.e. C lies to the left of AB

    If $\Delta > 0$ then CW   i.e. C lies to the right of AB

Hence the predicate can be expressed as

```
Bool COLLINEAR (A, B, C){
    If (Δ == 0) then
    return true
    Else
    return false
    }
```

Point–line predicates to find whether a point is to the left of a line or to the right of a line or on the line

Consider a point P (x, y) and the directed line segment AB, A (x_1, y_1) and B (x_2, y_2). Testing whether the point P lies to the left of AB is equivalent to testing whether the points P, A and B form a counter-clockwise (CCW) orientation in the ABP plane or not.

```
    Bool CCW (A, B, P) {
    If (Δ > 0) then
    return true
    Else return false
    }
```

Predicate to find line–line or segment–segment intersection (proper intersection, junction intersection, overlapping or coincident)

Testing whether the directed segment AB and CD intersect properly and exactly at one point can be done by evaluating the Boolean predicate INTERSECTPROP given below.

Bool INTERSECTPROP (A, B, C, D) {

    If (Collinear (A, B, C) || Collinear (A, B, D) ||
    Collinear(C, D, A) || Collinear (C, D, B)) then
    return false
Else
    return Xor (CCW(A, B, C), CCW (A, B, D)) &&
        Xor (CCW(C, D, A), CCW(C, D, B));
}

Predicate to find a point inside triangle, circle

To test whether a point P (x, y) lies inside a triangle constituted by points A (x_1, y_1), B (x_2, y_2) and C (x_3, y_3), imagine traversing from A to B, B to C and C to A, looking at point P. If the point P lies to the left of the viewer all the time or to the right of the viewer all the time, then P lies inside triangle ABC. One has to test whether the area subtended by triangle PAB, PBC, PCA are all positive. If all the angles are positive, then P lies inside triangle ABC else it lies outside.

POINT_INSIDE_TRIANGLE (P, A, B, C)
{

    If (((area PAB) > 0) && ((area PBC) > 0) && ((area (PCA) > 0)) ||
    ((PAB) > 0) && ((area PBC) > 0) && ((area (PCA) > 0)))
    then
    P lies inside triangle ABC
    Else P lies outside ABC
    }

In order to test whether a point P (x, y) lies inside a circle passing through points A (x_1, y_1), B (x_2, y_2) and C (x_3, y_3), one has to test the positive-ness of the determinant given below.

$$InCircle(A, B, C, P) = \det \begin{pmatrix} x_1 & y_1 & x_1^2 + y_1^2 & 1 \\ x_2 & y_2 & x_2^2 + y_2^2 & 1 \\ x_3 & y_3 & x_3^2 + y_3^2 & 1 \\ x^2 & y^2 & x^2 + y^2 & 1 \end{pmatrix}$$

Primitive to compute the area of a simple polygon

Consider a simple polygon with vertices $V(x_1, y_1)$, $V(x_2, y_2)$, $V(x_3, y_3)$ ....$V(x_n, y_n)$, where none of the segments bounding the polygon intersect with each other. Twice the area of the polygon P is given by

$$2\,A\,(P) = \sum_{i=1}^{i=n} \left(x_i y_{i+1} - y_i x_{i+1} \right)$$

9.5.2 Computation of Convex Hull

The convex hull problem in computational geometry revolves around computing the smallest convex shape that surrounds a given set of objects. A convex hull can be imagined as the shape of an elastic membrane snapped tightly around the object set.

In mathematical language, the problem can be stated as: To find the convex set of a set of points $P: = \{p_1, p_2, p_3, p_4.....p_n\}$ in a plane i.e. the convex hull of the set of points such that any segment joining two points of the set lies within the polygon set. There are a number of algorithms existing today to compute the convex hull of a set of points. One that is discussed here is the Graham's scan algorithm.

INPUT: A set of *n* points $P: = \{p_1, p_2, p_3, p_4...p_n\}$ in a plane
STEP 1   Find the right-most lowest point from the set (sort in *x* and then in *y*)
STEP 2   Sort points angularly about the right-most lowest point and store them in increasing order.
STEP 3   Push them in reverse order in a stack S
STEP 4   While ($i < n$) do
    4.1   If p_i is strictly left of the first two points of the points of the stack i.e. CCW(p_i, S(top −1), S(top)) then PUSH point p_i into the stack and increment *i*
    4.2   Otherwise POP S(*i*)
STEP 5   POP S

OUTPUT: The contents of S as the convex hull

9.5.3 Computation of Delaunay Triangulation

Given a set of points $P: = \{p_1, p_2, p_3, p_4.....p_n\}$ in space, is it possible to efficiently describe the relationship between these points? Perhaps the first data structure to do so is the Delaunay triangulation. The Delaunay triangulation of P is defined as the set of all triangles (p_i, p_j, p_k) such that, the circle passing through (p_i, p_j, p_k) does not contain any other point from P in its interior (this is known as empty circum-circle criteria). The graph generated by a Delaunay triangulation is known as the Delaunay graph (DG). An example of a DG is given in Fig. 9.5 (a).

A number of algorithms has been proposed by researchers on Delaunay triangulation. The run time complexity of these algorithms depends upon the number of points in the domain, how these points are distributed, the data structure used and the method of implementation.

The method of implementation of these algorithms can be grouped under four principal types, namely (1) divide-and-conquer, (2) random incremental insertion, (3) triangulation growth and (4) convex hull insertion algorithm.

The algorithm described below exploits the fact that an *n*-dimensional Delaunay triangulation can be deduced from a computation of the convex hull in *n* + 1 dimensions. The algorithms in the followings are algorithms for computing 3D convex hulls. Using the 3D points with coordinates (x, y, x^2+y^2), Delaunay triangulation of the set of 2D points with coordinates (x, y) is just the bottom part of the convex hull.

INPUT: Set of points P: = $\{p_1, p_2, p_3, p_4.....p_n\}$

STEP 1   Read the point set P into an array and y_n,
    where n = 1, 2... maximum number of points in P

STEP 2   For n =1... maximum no of points,
    Compute $z_n = x_n {}^\wedge 2 + y_n {}^\wedge 2$

STEP 3   For each triple (p_i, p_j, p_k)
    For $(i = 0; i < n - 2; i++)$
    For $(j = i + 1; j < n; j++)$
    For $(k = i + 1; k < n; k++)$
    If $(j != k)$ {

3.1:   Compute normal to triangle (p_i, p_j, p_k)
    $x_n = (y_j - y_i) \times (z_k - z_i) - (y_k - y_i) \times (z_j - z_i);$
    $y_n = (x_k - x_i) \times (z_j - z_i) - (x_j - x_i) \times (z_k - z_i);$
    $z_n = (x_j - x_i) \times (y_k - y_i) - (x_k - x_i) \times (y_j - y_i);$

STEP 4   Examine faces on the bottom of paraboloid due to (p_i, p_j, p_k) whose $z_n < 0$
    If $(flag = (z_n < 0))$

STEP 5

For all other points m does not fall below the paraboloid
    for $(m = 0; m < n; m++)$
    /* Check if m above (i,j,k). */
    flag = flag &&
        $((x_m - x_i) \times x_n +$
        $(y_m - y_i) \times y_n +$
        $(z_m - z_i) \times z_n <= 0);$

STEP 6   If no other points from P fall below the paraboloid due to (p_i, p_j, p_k)
    Then (p_i, p_j, p_k) is Delaunay conferment
    Add (p_i, p_j, p_k) to the output list
    If (flag) {
    printf("z = %10 d; lower face indices: % d, %d, % d\n", z_n, i, j, k);
    fprintf(f_p,"% f, % f, % f % f % f % f\n", x_i, y_i, x_j, y_j, x_k, y_k);
    }

STEP 7   Stop

OUTPUT: Set of three-tuple (p_i, p_j, p_k) forming the Delaunay triangulation

9.5.4 Computation of Voronoi Polygon

The Voronoi diagram of a point set P: = {p_1, p_2, p_3, p_4.....p_n} of unique random points in *n*-dimension space is defined as the set of sub-divisions D_i such that

$$D_i = \{x: I(x - p_i) \ I <= I(x - p_j)I \text{ for all } i, j, \text{ where } i \neq j\}$$

The collection of all such D_i is called the Voronoi tessellation of the point set P. The Voronoi region of a point consists of the points in the plane for which this point is the closest point of the set. The polygonal decomposition of the region represented by the point set is conforming.

Another definition of Voronoi obtained from the website http://mathworld.wolfram.com is given below:

'The partitioning of a plane with points into convex polygons such that each polygon contains exactly one generating point and every point in a given polygon is closer to its generating point than to any other is called a Voronoi diagram. A Voronoi diagram is sometimes also known as a Dirichlet tessellation. The cells are called Dirichlet regions, Thiessen polytopes, or Voronoi polygons.'

Voronoi tessellation of a surface is a dual geometrical structure of a Delaunay triangulation. A Voronoi tessellation can be obtained from the Delaunay triangulation by joining the meridians of three sides of each triangle. Thus both these structures share many common properties.

The outer edge of the Delaunay triangulation is the convex hull of the region. The Delaunay triangulation maximizes the minimum angle. Figure 9.5(a) shows the Delaunay triangulation of a set of points while Fig. 9.5(b) depicts the Voronoi polygon of the same set of points. To appreciate the dual geometric relationship of the two geometrical structures, Fig. 9.6 shows how a Voronoi diagram can be derived from Delaunay triangulations.

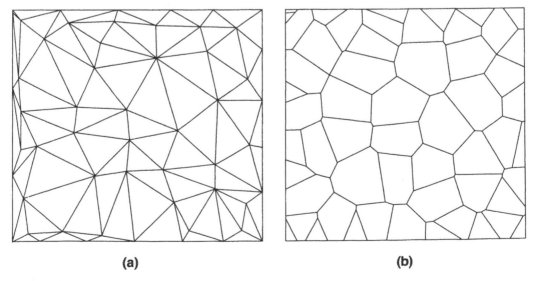

(a)   **(b)**

Fig. 9.5 (a) Delaunay triangulation of a set of points in 2D;
(b) Voronoi diagram of the same set of points

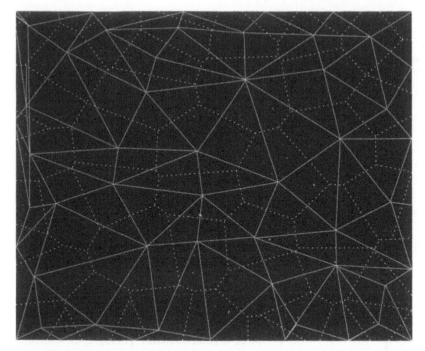

Fig. 9.6 Voronoi diagram as a dual geometry computed over a TIN

9.6 Application of CGAL Resources in GIS

From the foregoing sections it is clear that CG has the capability to abstract problems of different applied areas such as computer graphics, computer vision, robotics, CAD, GIS, scientific computing, spatial database design and query etc. Therefore CGAL is being extensively used in visualization and performing analysis of spatial data in GIS. Table 9.4 enlists some of the important applications where CGAL is put into use in GIS.

Table 9.4 CGAL resources used in GIS

Computational geometry resource	Application in GIS
Convex hulls	To find the minimum area, maximum area and optimum area occupied by a scattered set of points representing geographical point features
Line–line intersection	To find the intersection points of communication features e.g. road, rail, bridges etc.
Triangulation and shortest paths	Triangulating a monotone polygon divides the planar reason to a unique set of triangles which has following applications in GIS:

(continues…

Table 9.4 continued)

Computational geometry resource	Application in GIS
Range searching	
Point location	Point inside a polygon, triangle, volume etc. To find the known terrain feature within an geographical area
Voronoi diagrams	To find a suitable/optimum area for locating or installing a facility e.g. fire station, communication center, information center etc.
	Generation of a spatial index of a set of geometric objects. A particularly important geometric data structure in geographic analysis is the Voronoi diagram which has been used to identify regions of influence of clans and other population centers, model plant and animal competition, piece together satellite photographs, estimate ore reserves, perform marketing analysis, and estimate rainfall.
	Voronoi diagrams also help in determination of areas covered by a mobile telephone transmitter, determination of high-water marks in rivers, determination of noise zones along roads, drawing voltage maps, and other more conventional cartographic tasks.
Delaunay triangulation	To make an optimum representation / tessellation of a surface represented by a set of points in 2D/3D. To create a triangular irregular network (TIN) of points to give a representation of the surface under consideration. It is used in 3D digital elevation models
	3D DT for shape reconstruction of 3D geographic objects such as aquifers, ocean currents, and weather fronts.Generation of spatial index of a set of geometric objects.

9.7 Spatial Query and CGAL

Basically, a query is a mechanism used to retrieve relevant record or records (information) from an information system. It is also a mechanism by which the user obtains services from database-centric information systems. In the case of a geo-spatial database, the scope and usage of query is quite broad. The results of a spatial query can be textual, graphical or a visual pattern unlike traditional information systems where the result is predominantly a textual record or set of records. In GIS the query can be built through an interactive graphical query builder. The results of the queries from a GIS are graphically presented to the user and this is what makes the GIS a popular information system.

A query works by filtering a group of records from the entire database. Hence they are sometimes referred to as filters. When a user wishes to retrieve particular information or a set of information, he/she formulates a thought process expressed in the form of a query constituting of various predicates and conjectures.

If the query can be expressed completely as a predicate then it is a structured query, whereas if the filter or criteria cannot be precisely expressed through a predicate, it is known as a fuzzy query. Fuzzy query expresses the imprecise criteria in the thought process of a user through a mathematical expression involving probability.

There are different types of queries:

A simple query has one predicate or one criteria expressed in the form of a logical or mathematical expression.

A complex query builds a union or intersection of predicates through logical connectors thereby combining various criteria.

A hybrid query is a mixture of diverse criteria expressed through mathematical and logical connectors.

An iterative query is one where the result set of the query is an input to itself in successive steps.

A fuzzy query is one where the filter criteria are imprecise and expressed through fuzzy functions.

An unstructured query is one where the query cannot be expressed through a structured query language (SQL). It can be formulated by a combination of target data, geometric predicate and Standard English jargon.

A GIS system is popular depending upon the query builder or mechanism through which the user can formulate its queries, view and analyze its results. Modern GIS provides GUIs integrated through text, voice and event-driven devices to take the user inputs and formulate the query criteria. The results are graphically highlighted and depicted. In fact because of the ease and unique way of handling the query mechanism in geo-spatial database, its popularity and usage are increasing day by day. Other information systems and RDBMS are adapting and following the trends of spatio-temporal query methods of GIS.

The various types of queries a GIS can handle are given in Table 9.5.

Table 9.5 Different types of queries in GIS

	Type of query	Explanation
1	Spatial query	Query expressed through the location information (x, y, z) coordinates expecting to find what the object is i.e. given where it is, find what it is?
2	Spatio-temporal query	Query criteria expressed with the (x, y, z, t) coordinates and seeking an answer of what the object is i.e. what is the object which is at (x, y, z) during time t or which occurred during time period t.
3	Non-spatial query or attribute query	Given a particular attribute regarding the object, seeking to find where the object is i.e. given what it is, find where it is?
4	Buffer query	Buffer is a unique mechanism to filter the objects around an object meeting a filter criterion for e.g. finding all the objects along the corridor 100 meters on either side of the National highway.
5	Hybrid query	It is a mix and match of spatio-temporal and non-spatial criteria to formulate the filter.
6	Iterative query	The result set of one query is input for the next query or the query itself with varying criteria
7	Structured queries	Queries that follow the structured pattern of formation and can be formulated using SQL
8	Unstructured queries	Where the query criteria is not known *a priori* and need to be formulated from partial answers.
9	Fuzzy query	When the query criteria cannot be expressed as a geometric predicate i.e. when the criteria are probabilistic then they are modeled as fuzzy set criteria. The queries, which use fuzzy criteria, are known as fuzzy queries.

Unlike normal databases where the query formulation or the criteria solely depends upon the table names and fields of the database, spatio-temporal query takes into account the position information, time of occurrence or time period of the event in addition to the attributes describing the object. The process of establishing relationships among spatial objects is known as topology building. Topology gives the geometrical association among spatially arranged objects. In GIS, the objects are categorized into four categories i.e. point objects, line objects, polygonal objects and volume objects. Hence the data is stored in geometric data structures such as point, line, polygon or volume. The filter criteria are expressed in terms of geometric predicates and primitives. The query thus formed is fired to spatially indexed databases. The result set is a geometric object or set of objects depicting earth features. Computational geometric algorithms play a crucial role in every steps of query building and analysis of geo-spatial data. A complete process flow of spatio-temporal analysis using CGAL is depicted in Fig. 9.7.

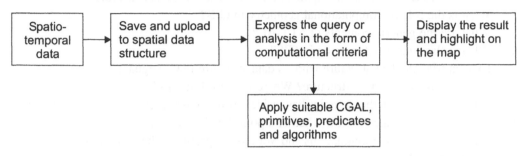

Fig. 9.7 Process flow of spatio-temporal analysis

References

[1.] Bowyer, A. Computing Dirichlet tessellation. *The Computer Journal* 24(2): 162–166. 1981.

[2.] Morino, Carlos. Efficient 2-D geometric operations. *C/C++ User Journal* 25–36 Nov. 1998.

[3.] Preparata, Franco P. and Shamos, Michael Ian. *Computational Geometry, An Introduction.* Springer-Verlag., 5th ed 1993.

[4.] Frank, A. U. Spatial concept, geometric data models, and geometric data structures. *Computers and Geosciences* 18:409–17.1992.

[5.] Aurenhammer, F. Voronoi diagrams: A survey of fundamental geometric data structure. *ACM Computer Survey* 23:345–405. 1991.

[6.] Peraire, J. et al. Adaptive re-meshing for compressible flow computations. *Journal of Computational Physics* 72: 449–462. 1987.

[7.] O' Rourke, J. *Art Gallery Theorems and Algorithms.* New York: Oxford University Press. 1987.

[8.] O' Rourke, J. *Computational Geometry using C.* New York: Cambridge University Press., 2nd ed, 1998.

[9.] Lo, S. H. A new mesh generation scheme for arbitrary planar domains. *International Journal for Numerical Methods in Engineering* 21: 1403–1426. 1985.

[10.] Lohner, R. Some useful data structures for generation of unstructured communications. *Applied Numerical Methods* 4(1): 123–135. 1988.

[11.] Lattuada, R. and J. F. Raper. Applications of 3D Delaunay triangulation algorithms in geo-scientific modeling In *Proceedings Third International Conference / Workshop on Integrating GIS and Environmental Modeling,* Santa Fe, Santa Barbara, NCGIA. CD, 21–26 January 1996.

[12.] de Berg, Mark, Marc Van Kreveld, Mark Overmars and Otfried Schwarzkopf. *Computational Geometry: Algorithms and Applications.* Heidelberg: Springer-Verlag. 1997.

Questions

1. What is CGAL? What are the application areas where CGAL is useful?
2. What is the role of computational geometry in GIS?
3. What are the generic contents of a CGAL?
4. What is the data structure used in GIS?
5. Explain triangle data structure with its data structure in C language.
6. What is a quad tree data structure? Where it is used in GIS?
7. How do we test whether a point lies inside a triangle using computational geometry?
8. Explain the predicate to show whether a point lies inside a circle.
9. What are Delaunay triangulations and Voronoi diagrams? Where are they used in GIS computations?
10. How is spatial query manifested using CGAL?

10

Spatial Interpolation

When an area is under study, it is difficult to survey or measure the spatial data on a continuous basis. Hence the spatial data domain processed by GIS is discontinuous and random in nature. For example, the demographic data of a study area is not uniform; they are geo-coded as discrete points. Often there is a need to determine the demographic value in locations which may not fall at the exact locations of the data points. To obtain the demography of the queried point, a surface is fitted to the surveyed data as a continuous phenomenon. This surface form is used to measure demography and predict or simulate the behavior of demographic phenomenon in a continuous manner. Basically, known values of object attributes are *interpolated and extrapolated* so that information can be spread upon the entire study area. This process of fitting a surface to a discrete set of surveyed data for computation of values at unsampled locations is often known as *spatial interpolation*.

This chapter introduces the important concepts of spatial interpolation. The definition of spatial interpolation is established in Section 10.2, followed by a discussion on various types of spatial interpolation techniques. Depending on the properties of the spatial interpolation function, they are categorized as local or global; exact or random; and isotropic or anisotropic. The choice of spatial interpolation for different applications of GIS is also discussed.

10.1 Introduction

Spatial interpolation is used for the generation of spatial data at unknown points within a scattered space of surveyed data. Spatial statistics gives a mathematical approximation to the spatial data from a range of known spatial data. Both spatial interpolation and statistical techniques together can be considered as a special set of analytical tools developed especially to process spatial data. Therefore, they are important techniques and are often studied as a separate branch of geographic information science known as *geo-statistics*. Spatial interpolation is commonly useful in modeling spatial distribution of phenomena such as temperature, precipitation, concentrations of chemicals, soil properties etc. It is especially useful in disaster management systems where direct measurement of the impact parameters cannot be done.

Measurement of certain parameters such as magnitude of seismic tremor, concentration of nuclear radiation or impact of a weapon in the zone of influence is done using spatial interpolation.

10.2 What is Spatial Interpolation?

An important concept behind spatial interpolation is the observation that points close together in space are more likely to have similar values than points far apart (Tobler's first law of geography).

Interpolation is derived from the Greek word *inter* meaning 'between' and *pole* meaning 'points' or 'nodes'. Any means of calculating a new point between two existing data points is known as interpolation. Spatial interpolation is defined as *'the procedure of estimating the value of properties at unsampled sites within the area covered by existing observations. Estimation outside the existing observations is known as extrapolation'*.

Therefore interpolation techniques are used when the sample space is 'up-sampled' to increase the resolution of the sample space or to assign values for an unsampled location from the sample locations. Sometimes 're-sampling' through interpolation is done to increase the concentration of the values on the sampled region. Sampling and re-sampling techniques are widely used in image processing to increase the optical resolution of the digital image or create a super-resolution image from different sub-sampled images.

In numerical analysis, *multivariate interpolation* or *spatial interpolation* is mainly interpolation of functions of more than one variable. The function to be interpolated is known at given points $(x_i, y_i, z_i, ...)$ and the interpolation problem consists of yielding values at arbitrary points $(x, y, z, ...)$.

There are many methods for doing this, many of which involve fitting some sort of function to the data and evaluating that function at the desired point. This does not exclude other means such as statistical methods for calculating interpolated data. In a two-dimensional space, a curve is used as the function; in 3D, a trend surface; and in 4D, a trend volume is generated out of the sampled points. This generated surface is often known as the model of the interpolation. The simplest form of interpolation is to take the mean average of x and y, two adjacent points, to find the mid-point. This will give the same result as the linear interpolation evaluated at the mid-point.

Mathematically, given a sequence of n distinct data points (x_k, y_k) called *nodes* and z_k, the values of the spatial entity at each node (x_k, y_k), where z can be the value of elevation, soil structure, chemical concentration etc., the data set containing the nodes and their values is known as sample space or survey space. Spatial interpolation is the search for a function f (interpolant function) such that for all k, $f(x_k, y_k) = z_k$ holds within the survey domain.

Hence for a given data set, modeling an interpolant function is important and is known as data fitting. Using this function, the value of the spatial object at any other unknown location (x_m, y_m) within the data domain can be computed. The argument of the interpolant function can be (x_k, y_k) in 2D, (x_k, y_k, z_k) in 3D or (x_k, y_k, z_k, t) when interpolation of the data in the time domain is required. Some examples of important interpolant functions are

- elevation: $z = f(x,y)$
- precipitation: $p_i = f_i(x,y)$; $I = 1,...,12$
- soil texture: $P = f(x,y,z)$, $P = \{S_d\}$, $d = 20, 80, ...$
- underground concentrations of chemicals: $w = f(x,y,z,t)$
- concentration of chemicals in water: $w = f(x,y,z,t)$

10.3 Usage of Spatial Interpolation in GIS

Spatial interpolation and extrapolation has a wide range of applications in GIS. Almost all GIS systems have different spatial interpolation functions implemented as a core analysis function of spatial data. Applications such as mining, disaster management systems and demographic systems, which have a strong spatial analysis component, make use of spatial interpolation because it can address the problem of spatial data unavailability under sampling. Spatial interpolation is a quick-fix solution for partial data coverage i.e., it fills the gaps between observations. These techniques are also used to generate trend surfaces from sample data so that the overall trend of the phenomena under consideration can be studied from the available data.

More specifically, spatial interpolation may be used

- To provide contours for displaying data graphically
- To calculate some property of the surface at an unmeasured point using known measurements
- To change the unit of comparison when using different data structures in different layers
- As an aid in the spatial decision-making process both in physical and human geography and in related disciplines such as mineral prospecting and hydrocarbon exploration

10.4 Different Types of Spatial Interpolation

Many types of spatial interpolation techniques have been developed over the years. These techniques have been developed to address specific applications which in turn call for handling of specific properties of the spatial data. The phenomenon which is to be modeled may use either the local property of the data or the property of the entire data space available; it may be stochastic or vary gradually over the trend surface. The interpolant function also depends upon the direction of the spatial variation in the data. Hence depending upon the property of the spatial phenomena which is to be modeled, the spatial interpolations can be categorized as

- a. Global or local interpolation
- b. Gradual or abrupt interpolation
- c. Exact or inexact interpolation
- d. Deterministic or stochastic interpolation

The nature of the spatial data and the application under consideration decides which type of interpolation function is used. The choice of an appropriate spatial interpolation method for a given data set depends upon the skill and expertise of the user of the GIS. The different types of spatial interpolation are explained in the following sections.

10.4.1 Global and Local Interpolation

Natural phenomena such as rain, cold, snow fall, temperature etc. are regional in nature and vary smoothly over a large area. This type of phenomena is known as regional phenomena. To model such spatial phenomena, only the spatial data of the entire region as a whole has to be taken into consideration and a single interpolation function is fitted to the entire region Hence a trend surface is fitted to the entire region based on a hypothesis framed for the global phenomena. The mathematical interpolating function thus obtained is not much affected by local variation of spatial data. Such an interpolating function which is derived from the data of the entire region is known as *global spatial interpolation.*

Often spatial phenomena such as traffic pattern, terrain undulation, mineral concentration etc. are purely local in nature. To model such phenomena, a mathematical function is designed which takes in a subset of the total observed points to predict the value of the unsampled point. Such spatial interpolation is called *local interpolation.* Given below is a comparison between local and global spatial interpolation.

Table 10.1

Sl. No.	Local methods	Global methods
1.	A single mathematical function is applied repeatedly to subsets of all the observed points	A single mathematical function is applied to all the points. Hence global interpolators determine a single function which is mapped across the entire study region
2.	A change in an input value only affects the result within the window	A change in one input value affects the entire map
3.	Local interpolation methods link regional surfaces into a composite surface, but they might not always be smooth	Global algorithms tend to produce smoother surfaces with less abrupt changes
4.	Some local interpolators may be extended to include a larger proportion of the data points in the set, thus making them, in a sense, global	Global algorithms are used when there is a known hypothesis about the form of the surface, e.g., a trend

Therefore the distinction between global and local interpolators is not a dichotomy but a continuum caused by how we use the surveyed data in generating the missing value.

10.4.2 Gradual and Abrupt Interpolators

Terrain surface and features have the peculiar characteristic of not being uniform all over. This variation cannot be modeled in a single mathematical function. Macroscopically, terrain surface is random in nature, varying abruptly. In a slightly lesser scale, terrain surface exhibits some uniformity such as sea surfaces, deserts or plains. A plain could be a vegetative or an urban area. To capture all these types of local variations and uniformity, a mixture of interpolating functions is used. An interpolating function which varies gradually and by and large continuously across edges will be able to model the gradual variation in spatial pattern whereas a highly discontinuous function which only depends upon the local pattern will be required to model the abrupt discontinuity of the terrain. Therefore both gradual and abrupt types of spatial interpolators are used to model spatial data. The table below gives a comparison of both types of spatial interpolators.

Table 10.2

Sl. No.	Gradual interpolator	Abrupt interpolator
1.	It produces a smooth surface between data points	It produces surfaces with a step-like appearance
2.	A typical example of a gradual interpolator is the distance weighted moving average interpolator.	The abrupt interpolator can be used to model abrupt changes in the terrain such as geologic faults and discontinuities in weather fronts
3.	It is appropriate for interpolating data with low local variability	It is appropriate for interpolating data with high local variability or data with discontinuities

A gradual interpolator usually produces an interpolated surface with gradual changes. However, if the number of points used in the moving average is reduced to a small number, or even one, there would be abrupt changes in the surface.

10.4.3 Exact and Inexact Interpolators

Sometimes the spatial data obtained is under-sampled or inadequate to predict values at missing points. In such a situation, the interpolation function needs to preserve the value at which the data points are sampled. This type of interpolating function is called an *exact interpolate*. Honoring data points is seen as an important feature of many spatial applications such as location of oil wells. Exact interpolation predicts a value at the point location that is the same as its known value. In other words, exact interpolation generates a surface that passes through the control points (Fig. 10.1 (a)).

Approximate interpolators are used where the data surface involves some amount of uncertainty. *Inexact interpolation* or approximate interpolation predicts a value at the point location that differs from its known value (Fig. 10.1 (b)). The table below gives a comparison of exact and inexact interpolations.

Table 10.3

Sl. No.	Exact interpolation	Inexact interpolation
1.	Exact interpolators honor the data points upon which the interpolation is based	Inexact interpolators do not honour all data points
2.	The surface passes through all points whose values are known	It utilizes the belief that in many data sets there are global trends which vary slowly, overlain by local fluctuations which vary rapidly and produce uncertainty (error) in the recorded values
3.	Proximal interpolators, b-splines and Kriging methods all honor the given data points	Since inexact interpolators do not honour all the input data points it is more appropriate when there is high degree of uncertainty about data points.
4.	It is appropriate for use when we know that we are dealing with accurate data	It is more appropriate when there is a high degree of uncertainty about data points.

Fig. 10.1 (a) Curve fitted with exact interpolation; (b) curve fitted with inexact interpolation

10.4.4 Stochastic and Deterministic Interpolators

The randomness of terrain surface cannot be modeled through deterministic functions; stochastic interpolators are better suited for this kind of problem. Generally stochastic methods incorporate the concept of randomness to spatial data. This is done by conceptualizing the interpolated surface as one of many surfaces that might have been observed, where all the values could have been produced from the known data points. An example of stochastic spatial interpolation is Krigging. Deterministic interpolation, on the other hand, uses sufficient knowledge about the distribution of spatial data on the surface to be modeled and builds a mathematical surface based on the data distribution.

10.5 Spatial Interpolation Methods used in GIS

Having understood the importance of spatial interpolation in analysis of spatial data and as a functional component of GIS, it is worthwhile to know some of the important spatial interpolation techniques often used in GIS. Most GIS packages offer a number of interpolation methods. The typical methods are:

1. Thiessen polygons or Dirichlet tessellation
2. Triangulated irregular networks (TINs)
3. Spatial moving average
4. Trend surfaces analysis

10.5.1 Thiessen (Voronoi) polygons

This method divides the sample surface to a set of interconnecting polygons based on the nearest neighbours. Thiessen polygon assumes that the value of unsampled locations are equal to the value of the nearest sampled points. First the sample space is tessellated to Voronoi regions as given below.

Let the sample space consist of a set of unique random points P: $\{p_1, p_2, p_3, p_4.....p_n\}$. Then the set of sub-divisions D_i such that

$$D_i = \{x: Ix - p_iI <= Ix - p_jI \text{ for all } i, j, \text{ where } i \neq j\}$$

generates a collection of Di which is called the Voronoi tessellation of the point set P. The Voronoi region of a point consists of the points in the plane for which this point is the closest point of the set. The polygonal decomposition of the region represented by the point set is a confor ming one.

Alternatively, Voronoi tessellation can be defined as *"the partitioning of a plane with points into convex polygons such that each polygon contains exactly one generating point and every point in a given polygon is closer to its generating point than to any other. A Voronoi diagram is sometimes also known as a Dirichlet tessellation. The cells are called Dirichlet regions, Thiessen polytopes, or Voronoi polygons"* (obtained from the website *mathworld*).

The fact that Voronoi tessellation of a surface is a dual geometrical structure of a Delaunay triangulation can be made use of to generate a Voronoi tessellation of the sample space. This is done by joining the meridians of the three sides of each triangle (Fig. 10. 2 (a) and (b)). It is a vector-based method. If the sample space consists of regularly spaced points, the Voronoi procedure produces a regular mesh. If the sample space consists of irregularly spaced points, it produces a network of irregularly shaped polygons as depicted in Fig.10.3(a) and Fig.10.3(b). Using Voronoi polygons is an exact, local, abrupt and deterministic spatial interpolation technique often used in sampling natural phenomena for hazard mapping, forest fire mapping or disease survey etc.

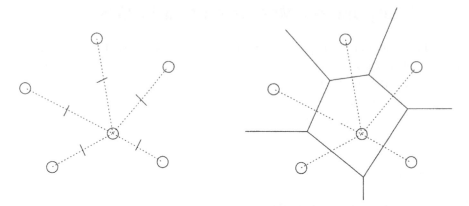

Fig. 10.2 (a) Medial axis of sampled locations, (b) Thiessian polygon of sampled locations

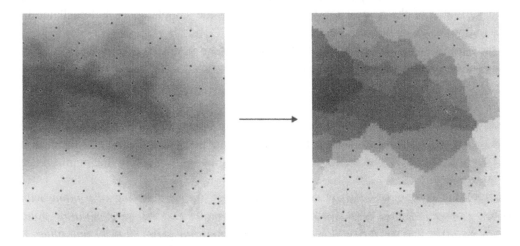

Fig. 10.3 (a) Source surface with sample points; (b) Thiessen polygons with sample points

10.5.2 Triangular Irregular Network

Triangular irregular network (TIN) is another popular vector-based spatial interpolation method often used—it is invariably present in almost all GIS softwares. It is generally used to create digital terrain models (DTMs). There are several procedures available today for generating TIN. Delaunay triangulation is an optimized and popular form of TIN well suitable for terrain modeling. A simple procedure for computing Delaunay triangulation was given in Chapter 9 (Section 9.5.3). A TIN connects adjacent data points (vertices) by lines (sides) to create a network of irregular triangles. First the TIN of the sampled locations is computed (Fig 10.4). Then the triangle containing the specified unsampled location (query point) is determined from the TIN using the predicate point-inside-tria ̃gle. Then the 3D distance between the data points along vertices and unsampled location is computed using trigonometry. The value of the unsampled location is the geometric mean of the values of the vertices of the triangle containing it (Fig 10.5).

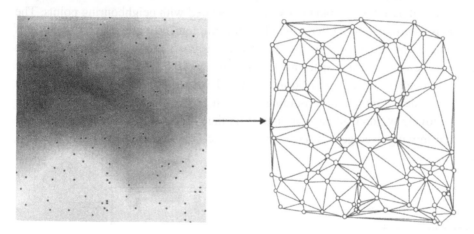

Fig. 10.4 (a) Source surface with sample points; (b) Resulting TIN

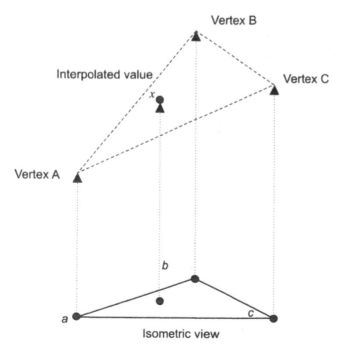

Fig. 10.5 Computation of geometric mean of the unsampled location using TIN

10.5.3 Spatial Moving Average (SMA)

Spatial moving average is a very popular interpolation technique used in GIS. It is based on the simple concept that the contribution of the sampled values at the unsampled location is a weighted average of their distance from the sampled locations. The value of the unsampled location is calculated based on the range of values associated with neighbouring points. The neighbourhood is determined by a filter or window. It is implemented as a moving average of proximal samples. The local samples are decided by a small moving window or circle convolving over the entire sample space as depicted in Fig. 10.6 (a) and Fig. 10.6 (b). The moving window decides which contributing sampled values at each instance are to be considered to compute the value of the unsampled location. Hence the size, shape and character of the filter or moving window all have an impact on the computed value at the unsampled location. Spatial moving average is a local, abrupt, exact and deterministic type of spatial interpolation.

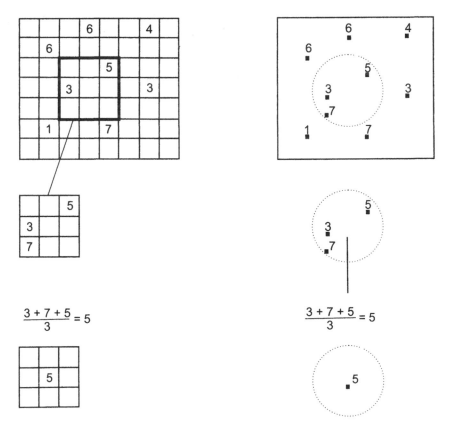

Fig. 10.6 (a) Rectangular; (b) Circular averaging window

10.5.4 Inverse Weighted Distance (IWD)

Inverse weighted distance (IDW) method is an important and popular spatial interpolation technique often used in many GIS applications such as computation of precipitation, temperature and other natural phenomena at unsampled locations. The principle used here is similar to that used in the SMA method except that here the weight is proportional to the inverse square of the distance of the sampled location. The governing equation of IWD is given in equation (10.1). It is often used in generating contours like isohyet, isobars, isotherms etc.

$$z_0 = \frac{\sum_{i=1}^{s} z_i \cdot \frac{1}{d_i^k}}{\sum_{i=1}^{s} \frac{1}{d_i^k}} \tag{10.1}$$

where

z_0 is the interpolated value at an unknown location.

z_i is the value of the ith sampled location.

d_i^k is the kth power of the distance between the ith sampled location and the unsampled location

If $k = 2$ then the IWD is known as inverse square distance interpolation. An example of an inverse square distance interpolation is given below.

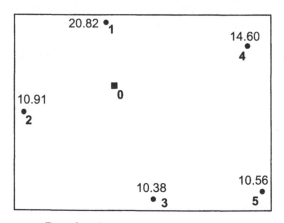

Between points	Distance
0,1	18.000
0,2	20.880
0,3	32.310
0,4	36.056
0,5	47.202

For $k = 2$

$$\Sigma z_i \frac{1}{d_i^2} = (20.820)\left(\frac{1}{18.000}\right)^2 + (10.910)\left(\frac{1}{18.000}\right)^2 + (10.380)\left(\frac{1}{32.310}\right)^2$$

$$+ (14.600)\left(\frac{1}{36.056}\right)^2 + (10.560)\left(\frac{1}{47.202}\right)^2$$

$$= 0.1152$$

$$\Sigma \frac{1}{d_i^2} = \left(\frac{1}{18.000}\right)^2 + \left(\frac{1}{18.000}\right)^2 + \left(\frac{1}{32.310}\right)^2 + \left(\frac{1}{36.056}\right)^2 + \left(\frac{1}{47.202}\right)^2$$

$$= 0.0076$$

$$z_0 = \frac{0.1152}{0.0076} = 15.158$$

10.5.5 Kriging

Kriging is a spatial interpolation technique developed by Georges Matheron, as the 'theory of regionalized variables', and D.G. Krige as an optimal method of interpolation for use in the mining industry. The basis of this technique is the rate at which the variance between points changes over space. The variance between points is expressed in the form of a graph known as a variogram which shows how the average difference between values at points changes with distance between the points.

Kriging assumes that the spatial variation of an attribute consists of three components: a spatially correlated component, representing the variation of the regionalized variable; a 'drift' or structure, representing a trend; and a random error term.

Interpolation using kriging involves three generic steps

a.   Computation of semi-variance from the sample space
b.   Computation of weights of the sample points with respect to the unsampled location
c.   Computation or interpolation of the value at the unsampled location

STEP 1   The measure of the degree of spatial correlation or dependence among the sampled points is the average semi-variance computed using equation (10.2).

$$\gamma(h) = \frac{1}{2n} \sum_{i=1}^{n} (z(x_i) - z(x_i + h))^2 \tag{10.2}$$

where

$z(x_i)$ = Value of the sampled location at x_i

$z(x_{i+h})$ = Value of the sampled location at an distance h from x_i

$\gamma(h)$ = Semi-variance of the sampled location due to n points

h = distance or lag between points

n = number of pairs of points separated by h

z = attribute value

STEP 2   The input data for Kriging is usually an irregularly spaced sample of points. To compute a variogram, we need to determine how variance increases with distance. It can be plotted using the sampled data and the above equation.

A variogram can be defined as a diagram or a 2D plot of the average difference in values of observed samples versus the distance between the observed samples. The vertical axis of the variogram is $E(z_i - z_j)2$, i.e. 'expectation' of the difference—the average difference in values at any two points, a distance d apart, where d (horizontal axis) is the distance between i and j.

STEP 3   Once the variogram has been developed, it is used to estimate distance weights for interpolation. Interpolated values are the sum of the weighted values of some number of sampled points where the weights depend on the distance between the interpolated and sampled points. A normal kriging uses a fitted semi-variogram directly in spatial interpolation. The general equation for estimating the z value at an unsampled point is given by equation (10.3)

$$Z_0 = \sum_{x=1}^{s} Z_x W_x \tag{10.3}$$

where z_0 is the estimated value, z_x are values at different points, W_x are weights associated with each sampled point, and s is the number of sampled points used in the estimation.

10.5.6 Trend Surfaces

Trend surface analysis is a regression method often used to fit a global function to a random set of sampled data. The interpolant function models the trend of the distribution of sampled data as a surface. Hence it is known as a trend surface. Trend surface analysis often uses a polynomial function to fit a least-squares surface to the sample data points, therefore it is also known as a least-square regression or polynomial regression.

The order of the polynomial decides the smoothness of the trend surface and is used to fit the surface. As the order of the polynomial increases, the surface being fitted becomes progressively more complex. Though it may be complex, a higher order polynomial will not always generate the most accurate surface. The accuracy of the trend surface is generally dependent upon the distribution of the sampled data. Lower the RMS error, the more closely the interpolated surface represents the input points. The most commonly used order of polynomials is 1 through 3 generating linear, quadratic and cubic interpolation respectively. First order (linear) trend surfaces, second order (quadratic) trend surfaces and third order (quadratic) surfaces are used to fit trend surfaces in many spatial applications. They are also used in image processing for interpolation of unknown samples.

10.6 Problems in Spatial Interpolations

Spatial interpolation is often treated as the outcome of intelligent guesswork for computing values at unsampled locations from known sampled values. In other words, the interpolated value is never exactly what exists on the ground. This is because of many factors such as:

1. Uncertainty in sampled input data to the interpolation process. This can be due to many factors such as the sample space consisting of too few data points; it being too

limited or clustered spatially. Sometimes even the uncertainty in the location and/or value of the sampling is imparted to the interpolation process.

2. Interpolation of unsampled points at the edge or boundary of the sampled domain often require more sample points outside the study area. Lack of this information leads to inaccuracies in the computation of values at unsampled locations near the boundary of the sample space. This is often known as the edge effect. To improve interpolation and avoid distortion at boundaries, sampling must be done in an area much beyond the sample space

3. Poor choice of interpolation techniques often result in error and uncertainty in computation of interpolated results.

Hence either inaccurate input samples or incorrect choice of interpolation techniques leads to erroneous computation of spatial data. Care should be taken to gather adequate samples from the study area with accurate measurements and survey equipment. Further the appropriate spatial interpolation technique should be applied to compute the value at unsampled locations.

References

[1.] Mitas, L. and H. Mitasova. Spatial interpolation. In *Geographical Information Systems: Principles, Techniques, Management and Applications* edited by P.Longley, M.F. Goodchild, D.J. Maguire and D.W.Rhind. Wiley. 1999.

[2.] Mitasova, H., L. Mitas, B.M. Brown, D.P. Gerdes and I. Kosinovsky. Modeling spatially and temporally distributed phenomena: New methods and tools for GRASS GIS. *International Journal of GIS* 9 (4), Special issue on integration of environmental modeling and GIS. 1995.

[3.] Burrough, P.A. *Principles of Geographical Information Systems for Land Resources Assessment*. Oxford: Clarendon Press. Chapter 8. 1986.

[4.] Davis, J.C. *Statistics and Data Analysis in Geology*. 2nd edition. New York: Wiley. 1986. (Also see the first edition published in 1973 for program listings.)

[5.] Hearn, D. and M.P. Baker. *Computer Graphics*. Englewood Cliffs, N.J.: Prentice-Hall Inc. 1986.

[6.] Jones, T.A., D.E. Hamilton and C.R. Johnson. *Contouring Geologic Surfaces with the Computer*. New York: Van Nostrand Reinhold. 1986.

[7.] Lam, N. 1983. Spatial interpolation methods: A review. *The American Cartographer* 10(2):129–149. 1983.

[8.] Mather, P.M. *Computational Methods of Multivariate Analysis in Physical Geography*. New York: Wiley. 1976.

[9.] Sampson, R.J. *Surface II*. revised edition. Lawrence, Kansas: Kansas Geological Survey. 1978.

[10.] Waters, N.M. Expert systems and systems of experts. Chapter 12 in *Geographical Systems and Systems of Geography: Essays in Honour of William Warntz* edited by W.J. Coffey. London, Ontario: Department of Geography, University of Western Ontario. 1988.

[11.] Some important websites on spatial interpolation are: http://www.geog.ubc.ca/courses/klink/gis.notes/ncgia/u40.html
http://www.geog.ubc.ca/courses/klink/gis.notes/ncgia/u41.html

Questions

1. Describe the different ways in which spatial interpolation algorithms can be classified.
2. How are the methods of interpolation categorized?
   Give examples of data types that require:
   - local or global interpolation
   - exact or approximate interpolation
   - gradual or abrupt interpolation
   - deterministic or stochastic interpolation
3. What categories does the Thiessen polygon method fall into?
   a. exact or approximate
   b. deterministic or stochastic
   c. gradual or abrupt
   d. local or global
4. What could be the possible use of Thessen polygon in GIS?
5. What categories does the TIN method fall into?
   - exact or approximate
   - deterministic or stochastic
   - gradual or abrupt
   - local or global
   What could it be used for?
6. What categories does the SMA method fall into?
   - exact or approximate
   - deterministic or stochastic
   - gradual or abrupt
   - local or global
   What could it be used for?

11

Spatial Data Accuracy

Accuracy of spatial data is an important issue in GIS. The confidence in and acceptability of a GIS by the user community largely depends upon the level of accuracy of the output range produced by the GIS. Besides the acceptance of the user community, the use of GIS by various applications also depends upon the degree of accuracy of the spatial output. Accuracy in general and spatial data accuracy in particular is a collective term for spatial data quality; absence of processing errors and uncertainty due to digital computation. The scale, resolution and precision of spatial data also play an important role in the level of spatial data accuracy.

In this chapter, the term 'spatial data' is collectively used for spatial data, intermediate processed data, processed information, spatial decisions, simulation results etc. Geographic data can be interpreted as geo-spatial data and spatio-temporal data whereby space and time are linked with the attributes to completely model geographic objects. Hence when accuracy of spatial data is being considered, the contributing factors can be spatial accuracy, temporal accuracy and attribute accuracy.

This chapter starts with the definition of accuracy and precision of spatial data. Then various components of spatial data accuracy are discussed. How different types of spatial data accuracy are categorized is illustrated. Further, the chapter also discusses the wayd of measuring and quantifying inaccuracy of spatial data. The answer to the question 'how to verify the accuracy of spatial data?' is illustrated with an example. The propagation of errors at various stages of spatial data processing is discussed.

11.1 Introduction

Accuracy of spatial data is affected cumulatively by many factors such as data capturing, data compilation, pre-processing, post-processing, cartographic transformation and geographic computation. Inaccuracy creeps into data during survey, aerial photography, scanning, digitization, registration, compilation, cartographic projection etc. The error imparted can be manual; or due to improper calibration of sensors; or instability of the platform that is capturing the spatial data.

Error also occurs in the next step of the GIS process—data conversion. Generally, in this step spatial data is converted from one format to another e.g. raster to vector format (R2V) using automatic or semi-automatic digitization. Importing spatial data from various sources to the native computing environment of the GIS or exporting to other environment sometimes imparts error.

Spatial data compilation aims at bringing together spatial data obtained from various sources and sensors. In a way, it is akin to spatial data fusion, where satellite images of various resolutions, surveyed vector maps and data obtained through GPS of the same area can be put together for better accuracy. But, often fusing spatial data from various sensors imparts error into the spatial data.

Further, a typical source of error is during computations, measurements, simulations and visualization of spatial data. This may be due to the malfunctioning of the computing logic applied to the data.

Another major point where error enters into spatial data is during the preparation of a spatial data repository i.e. spatial database design and management.

11.2 Definition of Spatial Data Accuracy

Highly accurate spatial data is data which is almost close to what occurs in nature and can be represented precisely in a digital environment. Formally, accuracy of spatial data can be defined as *'the closeness of results, computations or estimates of the data to its true values (or values accepted to be true)'*.

Since spatial data is usually a generalization of the real world, it is often difficult to identify a true value. Instead we work with values which are largely accepted to be true in our day-to-day use. e.g., in measuring the accuracy of a contour in a digital database, we compare the contour to one drawn on the source map, since the contour does not exist on the real earth surface as a closed line. Similarly the accuracy of the spatial database may have little relationship to the accuracy of products computed from the database. In such cases, the computed or measured value needs to be close to the value mostly accepted to be accurate.

It is impossible to represent infinite natural objects in a finite precision representation. Therefore while recording spatial objects in a finite precision digital environment such as a computer, the measurements pertaining to the object has to be scaled down, transformed and rounded or truncated to the upper bound of the computer's precision or word width. Therefore precision is defined as the number of decimal places or significant digits in a measurement of values representing the natural object.

Accuracy of spatial data is almost synonym to the precision of spatial data. But precession is not same as accuracy—a large number of significant digits do not necessarily indicate that the measurement is accurate. GIS works at high precision—sometimes the precision is much higher than the accuracy of the data itself. Besides which, the geo-spatial input itself is quite degenerate and complicated in nature. Hence conventional geometric computation is inadequate to process it and often yields wrong results. To overcome such problems, a set of robust computational geometric algorithms have been devised which computes results depending upon the sign and orientation of the geometric data rather than only its magnitude (Refer to Chapter 9).

11.3 Components of Spatial Data Quality

The quality of spatial data is affected by many factors influencing its collection, processing, computing and its representation in a digital environment. The various components that contribute to the quality of spatial data are:

- Positional accuracy
- Attribute accuracy
- Logical consistency
- Completeness
- Lineage

11.3.1 Positional Accuracy

Positional accuracy is defined as the closeness of the location information of spatial data to its true position. Conventionally, maps are roughly accurate to one line width or 0.5 mm which is equivalent to 12 meter of ground distance in a 1:24,000 map, or 125 m on 1:250,000 map. Positional accuracy is important in every application using spatial data, be it locating the object of interest in a map or precisely guiding a missile to its target location. In mission critical applications such as the landing of an aircraft or the guiding of missiles, an inaccuracy of even a few meters will lead to disaster. Hence ensuring positional accuracy in spatial data is very important; positional accuracy is tested in many ways before the data is put into use for various applications. The logical question arises 'how do we test positional accuracy?'

There are many ways of testing and verifying positional accuracy in a spatial data set, namely:

a. Using an independent source of higher accuracy for verification
b. Using maps of larger scale to verify the positions of objects in lower scale
c. Using the global positioning system (GPS) for ground truth validation
d. Using raw survey data as a supplement
e. Using internal evidence generated by the GIS system to supplement and validate the data.

For example, we can check the accuracy of the vector data generated by digitizing software using the statistics of the digitized data. The completeness is ascertained by comparing the number of geometrical figures generated by the digitizing software with the total number of surveyed features. Unclosed polygons, lines which overshoot or undershoot at junctions are indications of digitization anomaly. The sizes of gaps, overshoots and undershoots may be used as a measure of positional accuracy.

Positional data error can be directly measured using geodetic control and GPS. The most accurate of absolute positional data is obtained using the geodetic control network—a series of points whose positions are known with high precision. Global positioning systems are also a powerful way of augmenting the geodetic network. Often most positional data is derived from air photos. Aerial photography and satellite imagery are rich sources of air photos. The accuracy of these data depends on the establishment of good control points using GPS and geodetic control network. Spatial data from remote sensing is more difficult to position accurately because of the size of each pixel.

11.3.2 Attribute Accuracy

Inaccuracy due to inconsistency of attributes of the spatial data objects occurs when the attribute of the spatial object (point/line/polygon) is different from that of the attribute of the actual object in the ground.

Attribute accuracy is defined as the closeness of attribute values to their true values. Generally, location does not change with time, whereas attributes of spatial objects often change. For example, the position of a building, by and large, remains unchanged but the height and number of floors may increase because of construction, thus changing its attribute. Therefore, attribute accuracy of spatial objects must be analyzed in different ways depending on the nature of the data. Geographic attributes are classified as nominal, ordinal, interval, ratio and cyclic (Appendix C, Table 7). Attributes can be continuous or discrete. For continuous attributes (surfaces) such as on a digital elevation model (DEM) or triangular irregular network (TIN), accuracy is expressed as a measurement error e.g., elevation accurate up to 1m, or percentage error such as accurate to 90% etc. This error information is generally attached as an appendix to the main data or as meta information with the original data.

For categorical attributes such as classified polygons, the categories must be appropriate, sufficiently detailed and defined. Classification of remote sensed data is often done through automated software classifiers (typically artificial neural network or fuzzy classifiers) which use prior knowledge obtained through training to classify the data into different physical types. But however efficient the classifier may be, there is a margin of error which results in misclassification of objects e.g., a polygon may be classified as a water body (A type), when it should have been a marshy land (B type). Similarly a pixel classified as vegetation may actually be a building on the ground etc. This may be because the polygon was heterogeneous with 70% of the pixels of vegetation type and 30% of water. In such cases, the classifier may be confused as to which type it should classify the object. Sometimes, the very definition of a particular class is quite fuzzy and the classes A and B may not be well-defined, hence the software may not be able to clearly identify the object as A or B e.g. soils classifications are typically fuzzy. Here, at the center of the polygon, the object clearly belongs to class A, but more to class B at the edges. In such circumstances, the classification logic of the software inadvertently makes a mistake and imparts an attribute classification error.

The question in this scenario is: 'How to test attribute accuracy?'

There are many methods to check and measure attribute accuracy of spatial data. Ground truth is an expensive and time-consuming process and involves a lot of human-intensive work. Measuring the κ coefficient from the inconsistency matrix is a more practical method for checking the attribute inconsistency of a spatial data. It involves

    a. Preparation of the misclassification matrix from the spatial data
    b. Computation of omission and commission
    c. Computation of the κ coefficient

STEP 1   Prepare a misclassification matrix

From the classified data, take a number of randomly chosen points. Determine the class according to the database and then determine the class in the field by ground check. Suppose the classes are tagged as A (water body), B (vegetation) , C (building) and D (rocky area) in the database then prepare the misclassification matrix as a square matrix of classes as on ground vs. classes as classified in the database in columns and rows respectively (Table 11.1)

Table 11.1 Misclassification matrix of 4 classes

Class on ground → Class in database ↓	A	B	C	D	Total
A	4	4	2	3	13
B	1	7	1	3	12
C	5	2	4	3	14
D	6	5	1	6	18
Total	16	18	8	15	57

Ideally, all points must lie on the diagonal of the matrix and off diagonal elements should be zero—indicating that the same class was observed on the ground as is classified and recorded in the database by the classification software.

An error of omission occurs when a point's class on the ground is incorrectly recorded in the database. In the above misclassification matrix, the number of class B points incorrectly recorded is the sum of (column B, row A); (column B, row C) and (column B, row D), i.e. the number of points that are B on the ground but something else in the database. It can also be calculated as the column sum less the diagonal cell, which in this case = 18 − 7 = 11

An error of commission occurs when the class recorded in the database does not exist on the ground, e.g., the number of errors of commission for class A is the sum of (row A, column B); (row A, column C) and (row A, column D), i.e. the points falsely recorded as A in the database. It can also be calculated as the row sum less the diagonal cell which in this case = 13 − 4 = 9

STEP 2   How to summarize the misclassification matrix?

The percentage of cases correctly classified is often used as a measure of accuracy of the spatial data. This is the percentage of cases located in the diagonal cells of the matrix, called the kappa (κ) coefficient (equation (11.1)).

The κ index adjusts for this by subtracting the number expected by chance. The number expected by chance in each diagonal cell is found by multiplying the appropriate row and column totals and dividing by the total number of cases. Therefore kappa is 1 for a perfectly accurate data (all N cases are on the diagonal of the misclassification matrix). If κ is below one then the spatial data is not accurate. The following example will further elucidate the computation of the κ coefficient.

STEP 3 Calculating the kappa index

The kappa index (κ) is given by

$$\kappa = \frac{d-q}{N-q} \tag{11.1}$$

where, d is the number of cases in the diagonal cells

         q is the number of cases expected to be in diagonal cells by chance

         N is the total number of cases

In the above matrix, $q = \dfrac{n_{row} \times n_{column}}{N}$

Computing for each class

A: $\dfrac{13 \times 16}{57} = 3.65$

B: $\dfrac{12 \times 18}{57} = 3.79$

C: $\dfrac{14 \times 8}{57} = 1.96$

D: $\dfrac{18 \times 15}{57} = 4.74$

Total $= 14.14$

Therefore $\kappa = \dfrac{21 - 14.14}{57 - 14.14} = 0.16$

11.3.3 Logical Consistency

Logical consistency refers to the internal consistency of the data structure. In particular, it emphasizes topological consistency.

The questions that can be asked to ascertain the accuracy of the spatial database are:

a. Is the database consistent with its definitions i.e. schema and table structure?

b. If there are polygons, are they closed?

c. Is there exactly one label within each polygon?

d. Are there nodes wherever arcs cross, or do arcs sometimes cross without forming nodes?

Normalization is the process of efficiently organizing data in a database. There are two goals of the normalization process:

a. eliminating redundant data (for example, storing the same data in more than one table) and

b. ensuring data dependencies make sense (only storing related data in a table).

Both of these are worthy goals as they reduce the amount of space a database consumes and ensure that data is logically stored. This holds true even for spatial databases. The database community has developed a series of guidelines for ensuring that databases are normalized. These are referred to as normal forms and are numbered from one (the lowest form of normalization, referred to as the first normal form or 1NF) through to five (fifth normal form or 5NF). In practical applications, you will often see 1NF, 2NF, and 3NF along with the occasional 4NF. Fifth normal form is very rarely seen. Hence to achieve logical consistency, the spatial database needs to be normalized to at least up to the 4th normal form.

11.3.4 Completeness

Completeness concerns the degree to which the data exhausts the universe of possible items. The question asked here would be: 'Are all possible objects included within the database?' A positive answer to this question will determine the completeness of the spatial data. A negative answer results in partial spatial data which leads to partial representation of the domain of discourse. This in turn leads to inadequate spatial information. Hence the answer and outputs of spatial queries will be incomplete and incorrect. Therefore completeness of spatial data and correctness of spatial data are directly dependent.

11.3.5 Lineage

Lineage is the record of the data sources and of the operations which created the spatial database. It tries to answer some fundamental questions which affect the spatial data accuracy like:

• How was it digitized, and from what documents?

• When was the data collected?

• What agency collected the data?

• What steps were used to process the data?

• What is the precision of the computational results?

The answer to these questions is often a useful indicator of accuracy of spatial data.

11.4 Error in Database Creation

A major sub-system of GIS is the creation of a spatial database which involves design, development, normalization and optimization of the spatial data. Error is introduced at almost every step of the database creation. The process of creating a spatial database is different from that of the development of a normal database, but the fundamental principles are the same. The possible errors in the various steps of spatial database creation are explained in the following sections.

11.4.1 Data collection errors

These are errors due to survey, scan, aerial photography or digitization. Often when the original data is digitized, errors may be introduced due to human faults or software malfunction. The anomalies caused by digitization are explained in detail in Section 11.5.

Sometimes errors are also introduced because of the poor stability of the base material on which the map or aerial image is taken initially e.g. paper. Paper can shrink and stretch significantly (as much as 3%) with changes in humidity. This introduces error in the overall position and scale of the spatial data.

11.4.2 Errors due to coordinate transformation

Coordinate transformation introduces error, particularly if the projection of the original document is unknown, or if the source map has poor horizontal control. Errors in registration and control points affect the entire dataset. As is evident, map transformation always imparts errors to spatial data. The errors may be in terms of position, distance, direction or area. Often these errors are mutually exclusive and non-uniform across the entire data set. Hence depending on the transformation, the error in the map is judged and mentioned with the transformed spatial data as meta information such as error record of the digital elevation model data (DEM). Spatial accuracy information is also mentioned in products such as maps, navigation charts or elevation models etc.

11.4.3 Attribute errors

Attributes are usually obtained through a combination of field collection and interpretation. Attributes are also attached to a spatial object after due classification using automatic software classifiers. Often data is classified using false–positive or false–negative categories by the classifier. This results in mismatch of attributes in the database and what is actually there on the ground. Often the attribute categories used in interpretation may not be easy to verify in the field. Terms such as 'high', 'moderate', 'low', which are often used, are highly subjective. Attributes obtained from air photo interpretation or classified satellite images may have high error rates. In case of social data or demographic data, the major source of inaccuracy is undercounting e.g., in census, undercount rates can be as high as >10% in some areas and in some particular social groups.

11.4.4 Compilation errors

Spatial data are collected from various sources and agencies using different methods. They are also stored in different formats and can be of different scales. Hence there is a need to compile the data into a composite data set so that the spatial object can be modeled completely. Common practices in map compilation such as generalization to make the data symbolic, aggregation, thematic composition, format conversion and fusion of spatial, introduce what is known as compilation error.

11.5 Digitization Anomalies

One of the main sources of error in creation of spatial data is digitization. Digitization encodes manuscript lines or raster image lines as sets of (x,y) coordinate pairs. Resolution of coordinate data depends on the mode of digitizing, expertise of the person doing the digitization and the efficiency of the digitization software. There are two modes of digitization namely

    a. Point-mode

    b. Stream-mode.

In point-mode digitization, the software requires intelligence and knowledge of the line representation that will be needed for the spatial data under consideration. In stream-mode digitizing, the device automatically selects points on a distance or time parameter. Generally, a high density of coordinate pairs is selected.

Two types of errors normally occur in stream-mode digitizing. Digitization errors can be caused by human inadequacies—agitations as the operator's hand twitches and jerks while digitizing leads to three main types of anomalies. These specific digitization anomalies may be identified as spikes in digitization which manifest as overshoots and undershoots in the data. The switchback error is another anomaly which results in unclosed polygons thus causing topological error while defining area objects. Loops often result in tiny and elongated skinny polygons known as silver polygons. A skinny polygon cannot model a terrain object appropriately.

Besides these types of digitization errors, the operator often fails to assign the appropriate symbol to the digitized object such as colour, line or point style, line width, polygon fill pattern, and display. Due to these types of errors, an object may not be represented visually according to the legend used in the spatial data. This will lead to misinterpretation of the spatial data.

With the advent of software tools all these types of digitization anomalies can be isolated and removed. This forms a part of the internal quality process of organizations producing spatial data. For example, a common problem in point-mode digitizing is duplicate coordinate pairs which occur when the button is hit twice. The QC (quality check) software may notify and can eliminate duplicate coordinate pairs or flag it as errors. It is not difficult to follow a line to an accuracy equal to the line's width while digitization. Typical errors are in the range of 0.2–0.5 mm. A common test of digitizing accuracy is to compare the original line with its digitized and plotted version by overlaying the raster and vector data or by manual inspection of the plotted version of the digitized map. The common digitization anomalies and their rectification process are discussed in the following sections.

11.5.1 Overshoots

Overshoots while digitization manifest as short line segments (Fig. 11.1(a)) that incorrectly extend beyond the actual geographical objects. A selection query such as,

> LINES.Length < [<Threshold value>]

will isolate all the line segments which are shorter then the threshold value. The very short lines can be treated as overshoots or undershoots which can then be possibly removed.

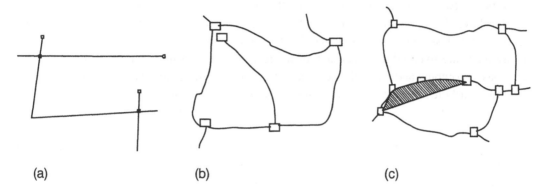

(a)   (b)   (c)

Fig. 11.1 (a) Overshoot/Undershoot; (b) Unclosed polygons; (c) Silver polygons

11.5.2 Unclosed Polygons

In vector data, objects containing a network of polygons, a gap between two lines that should have intersected result in unclosed polygons (Fig.11.1(b)). It may leave as a single polygon where two separate polygons should exist. Lines that fail to close a polygon can be found by querying

> If (Internal.LeftPoly == Internal.RightPoly) then unclosed polygon

because they have the same polygon on both sides.

11.5.3 Sliver Polygons

Sometimes the digitizer traces a line or boundary twice. Double-tracing polygon boundaries can create extraneous sliver polygons along the boundary of the two contiguous polygons. Silver polygons usually have a much smaller area than the main polygons, and are usually highly elongated (with a high compactness ratio). These are also called skinny polygons (hatched portion) (Fig.11.1(c)). We can use the following combined query on the area and compact ratio fields of the polygons:

> POLY. Area < [threshold value] {or/and} POLY. CompactRatio < [threshold value]

The result will select all sliver polygons in the data set.

11.6 Error Analysis and Sensitivity Analysis

Analyzing the impact of error on the overall output is quite challenging. Several pertinent questions while analyzing error in GIS are

    a. What impact does error in each data layer have on the final result?

    b. Are errors in each layer independent or are they related?

    c. Will the effects of cascading error be complex?

    d. Do errors get worse, i.e. multiply? Or do errors cancel out?

Suppose two maps, each with a correctly classified percentage of 0.90, are overlaid. Studies have shown that the accuracy of the resulting map (percentage of points having both the overlaid classes) is little better than $0.90 \times 0.90 = 0.81$. But it is not always true. When many maps are overlaid, the accuracy of the resulting composite can be very poor

One has to compute and know the accuracy of the composite suitability index rather than the overlaid attributes themselves. For some types of operations, the accuracy of suitability is determined by the accuracy of the least accurate layer. This is true if reclassification and the 'AND' operator are used extensively, or if simple conditions are used based on inaccurate layers. In other cases, the accuracy of the result is significantly better than the accuracy of the least accurate layer. This is true if weighted addition is used, or if reclassification uses the 'OR' operator   e.g. suitability = 4 if $x_1 = A$ OR $x_2 = D$.

Sensitivity of the GIS is a measure of how the result and outputs of GIS change with a change in the input of GIS. Sensitivity can be defined as the response of the result (suitability, or the route location) to a unit change in one of the inputs. It will answer the following questions

    a. The data inputs: how much does the result change when data input changes

    b. The weights: How much does the result change when the weight given to a factor changes

Error in determining weights may be just as important as error in the database. It may be better to use the full observed range to test sensitivity i.e. response of the result to a change in one of the inputs from its minimum observed value to its maximum. One can also use sensitivity analysis to assess the effects of uncertainty in the data due to computation. Sensitivity provides a measure of the 'confidence interval' of the results. It may also refer to spatial resolutions and answer queries such as

    a. Would increasing resolution give a better result?

    b. Would cost of additional data collection at higher resolution be justified?

    c. Can we put a value on spatial resolution?

Questions

    1. Define spatial data accuracy. What are the components of spatial data accuracy?

    2. What is positional accuracy? How do we test and validate positional accuracy?

    3. What is attribute accuracy? How do we test attribute accuracy?

    4. What is kappa coefficient? How is it computed and interpreted?

    5. What are the different anomalies in designing spatial database?

    6. What are digitization anomalies? How do we isolate them in a database?

Plate 1

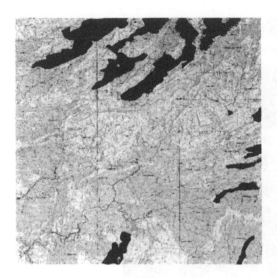

Fig. 5.5
Raster image/scanned image
of the map

Fig. 5.6
3D perspective view of the DEM with
raster image and vector layer draped on it

Fig. 5.7
Shaded relief view of the map

Plate 2

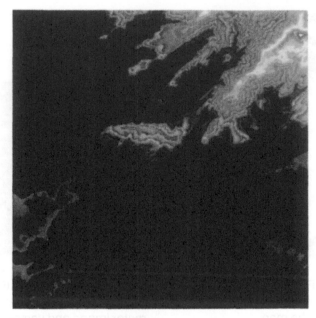

Fig. 5.8
Colour-coded elevation map

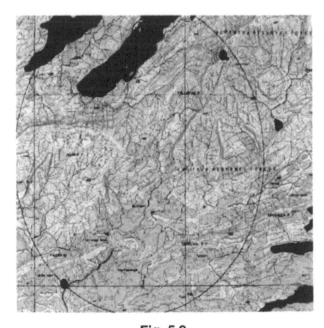

Fig. 5.9
Colour-coded elevation map Computing the
minimum and maximum height of a circular area

Plate 3

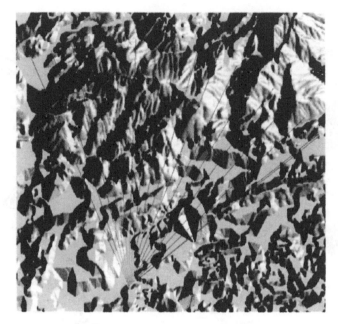

Fig. 5.10
Line of sight in the view cone
(red line: invisible; green line: visible)

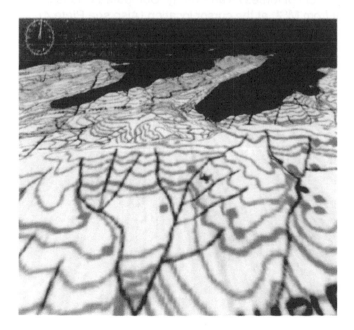

Fig. 5.11
Fly through visualization over the area with raster
image overlaid

Plate 4

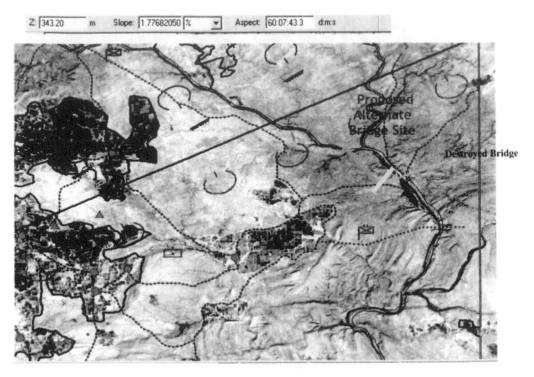

Fig. 8.14 Deployment of forces as an overlay; Computed values of slope, aspect and height from MSL at the cursor location (Also see Plate 4)

12

Evolution of Geographical Information Science

Geographical information science has established itself as a mature field of research and is the scientific basis of the rapidly evolving GIS technology. Research trends in GIS as indicated by scientific publications and papers presented at various international symposia addresses many scientific questions about analysis, visualization, computation and modeling of spatial and temporal data—many of which are beyond the technical domain of GIS technology as of today.

In this chapter we will review the progress and current state of research in GI science. Geographical information science is defined and its contributing scientific fields explored. The technologies associated with these scientific fields which are the driving aspects of GI science are also discussed.

This chapter has been organized into four sessions including the introduction. The second section tracks the major historical milestones that have impacted the developments in GI Sc. The third section discusses the allied scientific fields of GI Sc. The fourth section reviews the emerging trends of GI Sc. and discusses the possible direction of research in this field.

12.1 Introduction

The genesis of geographic information systems (GIS) can be traced back to many endeavors in applied research and development during the 1960s. Some of the important milestones are the formation of a GIS laboratory at the Harvard Laboratory for Computer Graphics and the development of Canada GIS. GIS as of today is based on a sound scientific basis known as Geographic information science (GI Sc.). Research in the area of GI Sc. has being carried out worldwide at several universities and scientific organizations over the past decades.

Human endeavor to explore the earth surface and hence the preparation of maps by putting the curved earth on a flat surface dates back to the 15th century. The early work of Gerardus Mercator (1512–1594), Johann H. Lambert (1728–1777) and many others laid the foundations for the modern-day scientific fields called geodesy and cartography. The mathematical contributions towards developing the theory for projecting the oblate spheroid earth surface on to a flat map were made by Gerardus Mercator and subsequently improved by Lambert. These techniques find their digital adaptation in modern-day GIS products.

Modern-day GIS is driven by a set of digital techniques called the geo-processing tool box. The geo-processing toolbox is a collection of software tools for many scientific techniques implemented through computational data structures and algorithms. The GI science which is at the heart of GIS manifests itself in this suite of computing techniques to efficiently organize spatio-temporal data and help in its analysis, measurement, visualization, modeling and interpretation.

The GI science as of today is a conglomeration of many scientific principles that are drawn from diverse disciplines such as cartography, computer graphics, spatial interpolation, computational geometry, graph theory, geo-statistics, database management, applied mathematics etc.

The collaborative nature of GIS has spurred the invention of many creative products which has caught the imagination of the user community in general and the scientific community in particular. Innovative applications like Google Earth, Sky and Planet have captured the imagination of novice users and drawn the attention of researchers of geographical information science. The spatial and dynamic nature of GIS data has also increased due to the phenomenal growth in a range of other technologies such as mobile communication and computing, global positioning systems, laser ranging, very high resolution remote sensing etc. While the emergence of GI science was driven by the application of GIS, there is now a global focus on developing standards to enable the proper utilization of the huge investments made by organizations in diverse commercial products and data formats.

12.2 Historical Perspective of GI Science

The initial evolution of GI science can be attributed to the need for accurate navigation, preparation of maps, and modeling of the earth surface. In a sense the genesis of GI Sc was due to the early work done in cartography and geodesy. As mentioned in the first section, the prominent cartographic theory which laid the foundation stone of modern digital cartography is the work done by Mercator on map transformation (equations 12.1). The Mercator map projection is still a major part of the map transformation tool box of every GIS.

$$x = R\,(\lambda - \lambda_0) \tag{12.1a}$$

$$y = R\ln\tan\,(\pi/4 + \phi/2) \tag{12.1b}$$

$$y = R\ln\text{arc }\tanh\,(\sin\phi) \tag{12.1c}$$

where, R = radius of earth in meters
ϕ = latitude in radians
λ = longitude in radians

Lambert later improved on these transformations, developing what is known today as Lambert's conformal conic (LCC) projection (equations 12.2).

$$x = a\,(\lambda - \lambda_0) \tag{12.2a}$$

$$y = a \ln\left[\tan\left(\frac{\pi}{4} + \frac{\phi}{2}\right)\left(\frac{1 - e\sin\phi}{1 + e\sin\phi}\right)^{e/2} \right] \tag{12.2b}$$

$$x = \rho(\sin\phi)$$

$$y = \rho_0 - \rho(\cos\phi)$$

where. $\theta = n(\lambda - \lambda_0)$.

$$n = \frac{(\ln m_1 \; \ln m_2)}{(\ln t_1 \; \ln t_2)}$$

$$m = \frac{\cos\phi}{(\ln e^2 \sin^2\phi)^{1/2}}$$

$$t = \frac{\tan\left(\dfrac{\pi}{4} + \dfrac{\phi}{2}\right)}{\left(\dfrac{1 - e\sin\phi}{1 + e\sin\phi}\right)^{e/2}}$$

and a = semi-major axis of the elliptical earth,
         e = eccentricity of the earth surface

The oblique aspect of the stereographic projection in its spherical form (equations 12.3) was derived in 1962. It has being developed and used for the projection of the moon surface, part of Mars surface and the images obtained by earth observation satellites.

$$\rho = 2R \tan\frac{1}{2}C \tag{12.3a}$$

$$\theta = \pi - Az = 180° - Az \tag{12.3b}$$

where, C = angular distance from the center of earth,
       Az = azimuth east of north and
       θ = polar coordinate east of south

The rectangular form of the above equation is given by equations 12.4

$$x = Rk \cos\phi \sin(\lambda - \lambda_0) \tag{12.4a}$$

$$y = Rk[\cos\phi_1 \sin\phi - \sin\phi_1 \cos\phi \cos(\lambda - \lambda_0) \tag{12.4b}$$

Over the past one decade, the progress in research and development in the GIS field is commendable. As mentioned before, this is substantiated by the ever-increasing number of research publications in the field of GI science in various international journals and symposiums. Introduction of a curriculum based on GI science by major academic institutes and industry is an indicator of the scientific and academic trends in GIS. Innovative applications involving spatial data processing, visualization and modeling are being developed and launched by various system developers. This rapid increase in GI science research has had great impact on the development of innovative GIS systems.

It has been established that GI science is the scientific basis of systems involving GIS. The significant historical developments which have influenced the progress of GI Sc have been put in a chronological order:

1963	Development of Canada Geographic Information Systems (CGIS) commences. Led by Roger Tomlinson to analyze Canada's national land inventory, it is considered as the first institutional effort to develop GIS (Tomlinson et al. 1976).
1964	The Harvard Laboratory for Computer Graphics and Spatial Analysis, Harvard University, USA is established by Howard Fisher.
1966	SYMAP, a mapping software developed by Howard Fisher, was used in a landscape-planning study of the peninsula, which was part of a GIS course at the Harvard Laboratory for Computer Graphics and Spatial Analysis. The educational and research program grew through the 60s, 70s and the 80s. Apart from the SYMAP, other Harvard packages, which were equally important in the developing field of GIS and spatial data analysis, were CALFORM (late 1960s), SYMVU (late 1960s), GRID (late 1960s), Polyvrt (early 1970s) and ODYSSEY (mid 1970s).
1969	Jack and Laura Dangermond founded ESRI, a privately held consulting group. ESRI's early research and development in cartographic data structure, specialized GIS software tools and creative applications set the stage for today's revolution in digital mapping. ESRI continues to set standards in the GIS industry. Its software is installed at more than 100,000 client sites worldwide and over 91 distributors.
1984	The National Center for Geographic Information and Analysis (NCGIA) was formed by the US National Science Foundation. The scope of core research topics were enunciated as a) Spatial analysis and spatial statistics b) Spatial relation and database structures c) Artificial intelligence and expert system d) Visualization e) Socio-economic and institutional issues Later the scope were refined to the topics under (a), (b) (d) and (e)
1985	The GPS (global positioning system) became operational. The development of GRASS (geographic resources analysis support system), a raster-based GIS program, started at the US Army Construction Engineering Research Laboratories. This was the beginning of the open GIS effort (GIS users would be able to freely exchange data over a range of systems and networks without having to worry about conversion of format or data types).

(continues...

Table continued)

1992	The term 'geographical information science', was first coined by Michael F. Goodchild. Based on his keynote address at the Fourth International Symposium on 'spatial data handling' at Zurich in 1990 (Goodchild 1992), and the European Conference on Geographical Information Systems at Brussels in 1991 (Goodchild 1991), a formal definition and scope of GI science was arrived at: GI science studies the fundamental issues arising from the creation, processing, and visualization of geographic information—we can see that the scope of GI science was restricted to only the processing and visualization of spatial data.
1995	The University Consortium of GIS (UCGIS) was formed (Mark and Bossler 1995). UCGIS is dedicated to the development of theories, methods, technologies and data for understanding GI processes. The definition of GI science evolved to 'the transformation of spatial data to useful information'.
1999	The National Science foundation of US gave a full definition of GI science as 'the basic research field that seeks to redefine geographic concepts and their use in the context of geographic information system'.

The scope of GIS science has increased many folds; so also its input domain (Table 12.1), processing capabilities and range of outputs.

Table 12.1 Input domain of GIS

Sl. No	Input data type	Source	Topology/Format
1	Raster scanned data	Scanner, unmanned aerial vehicle, oblique photography	Matrix of pixels with the header containing the boundary information, GeoTIFF, GIF, PCX, XWD, CIB, NITF, CADRG
2	Satellite image	Satellite	BIL, BIP, TIFF
3	Vector map	Field survey, digitizer output	DGN, DVD, DXF
4	Attribute data	Field survey, statistical observation, census data	Textual records binding several attribute fields stored in various RDBMS e.g. Oracle, Sybase, BDASE etc.
5	Elevation data	Sensors, GPS, DGPS LIDAR (hyper-spectral data)	Matrix of height values approximating the height of a particular grid of earth surface. DTED-0/1/2, DEM, NMEA, GRD, TIN
6	Marine navigation charts	Marine survey, coast and island survey, hydrographic and maritime survey organization	S57 / S56 electronic navigation charts, coast and island map data

(continues...

Table 12.1 continued)

Sl. No	Input data type	Source	Topology/Format
7	Ellipsoid parameters/ geodetic datum/ geo-referenced information/ coordinate system information	Geodetic survey, marine survey, government surveying organization	Semi-major axis, semi-minor axis, flattening/ eccentricity, origin of the coordinate center, the orientation of axis with respect to axis of earth centered—earth fixed reference frame.
8	Projection parameters	Geodetic survey organizations or agencies	Meta data or supporting data to the main spatial data, header information of the main file
9	Almanac and met data	Almanac tables	Time of sun rise, sun set, moon rise, moon set, weather information including day and night temperature and wind speed etc.

GIS has changed from an isolated information island to a collaborative platform for systems computing spatio-temporal data. The increased scope of GIS is changing the definition of GI science.

The definition of GI science as of today is 'the study of basic issues arising out of processing, visualization and modeling information and events pertaining to geo-spatial and spatio-temporal data'.

12.3 The Contributing Disciplines of GI Science

There is a pressing need to recognize and develop the role of GI science and delineate its allied technologies and science. This study will canalize the energies of the GIS research community to formally extend their intellectual curiosity and advance research in GI science. If development of GIS is guided by GI science., the question arises as to what are its sub-fields and how have they contributed to GIS. The sub-fields contributing to GI seience have been formally tabulated in Table 12.3 with the corresponding scientific principle leading to the software component in GIS.

With the advent of digital computation, traditional cartography, geodesy, geometry and geography have been transformed to their digital form as digital cartography, geodesy, computational geometry and computational geography respectively. The digital and computation form of these scientific fields is the basis of GIS. They provide a set of computation algorithms, robust and capable of computing large amount of spatial data. The key technology and sub-fields of GI science. which contribute to the progress of GI science are depicted in Fig. 12.1.

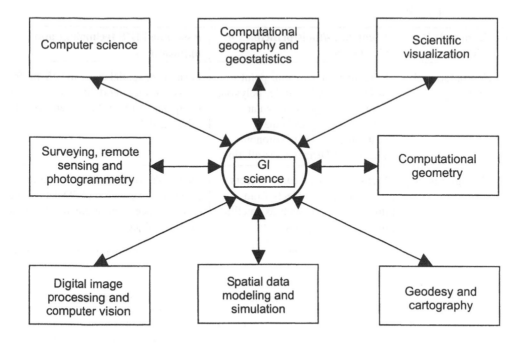

Fig. 12.1 Contributing technology and sub-fields in GI science

Table 12.2 Principle of GI science and corresponding technology in GIS

Sl. No	Branch of GI science	Algorithms/tools/principles in GI science	Associated GIS technology and applications
1	Digital cartography	Map transformation, coordinate transformation	Digital map preparation and mapping applications, 3D terrain visualization
2	Geodesy	Computation of horizontal and vertical geodetic datum of earth Modeling of earth shape and size. Computation and updation of major and minor axis of earth, eccentricity, flattening and curvature of earth surface at every point.Molondensky transformation, datum transformation	Modeling and study of earth surface
3	Computational geography	Algorithms to compute distance, direction, slope, aspect etc.	Computation of position, distance, height, direction, area, perimeter, volume, speed, slope, aspect, magnetic declination of a geographic feature

(continues...

Table 12.2 continued)

Sl. No	Branch of GI science	Algorithms/tools/principles in GI science	Associated GIS technology and applications
4	Computational geometry	Line–line intersection, point inside a triangle, circle, polygon, volume, triangular irregular network (TIN), Delaunay triangulation (DT), Voronoi/ Thessian / Dirichlet tessellation	Forming of spatial query in GIS, buffer query, surface modeling, and spatial interpolation of attributes pertaining to missing data.
5	Photogrammetry	Measurement of earth objects and man-made objects from images of earth such as satellite images, ortho-photos etc.	Ortho-photo generation, contour generation, generation of 3D data of the earth surface, measurement of height of earth objects.
6	Computer science	Databases, computational geometry, image processing, pattern recognition, spatial data mining, information science	Spatial data mining to predict patterns of hazards such as earth quakes, floods, spatial data fusion to obtain high quality spatial data.
7	Remote sensing	Sensing of earth surface remotely	Natural resource planning, natural disaster monitoring and warning.
8	Spatial statistics and spatial interpolation	Spatial interpolation such as Krigging, inverse distance weight of spatial pattern	Prediction of values and attributes of missing spatial data, computation
9	Science and technology traditionally dealing with geospatial and spatiotemporal data	Geology, geophysics, oceano- graphy, agriculture, biology, particularly ecology, bio- geography, environmental science, geography, sociology, political science, anthropology	Navigation in sea, air and ground, health science, demography, etc.

These contributing fields of GI science. have lots of scope in the way they can handle geospatial and spatio-temporal data and will eventually lead to algorithms and computing techniques which can handle large volume of spatial data and its degeneracy efficiently.

12.4 Evolution of GI Science

The initial phase of development of GIS technology was largely guided by scientific principles of cartography and geodesy. For example, the concept of map overlay techniques to represent the spatial patterns of events with a digital map in the background was developed by McHarg (McHarg 1969) and later put into operation by Berry in the year 1987 (Berry 1987).

The later phases of the subject were reverse processes where applications drove the development in technology trends. The development of the Canadian Geographical Information

System (CGIS) (Tomilison et al. 1976) is an example. CGIS was an application in search of technology. It sufficiently drove the research in the field of GIscience and lead to the development of new technologies such as the map scanner.

Of late the applications and technologies of GIS have far out-paced the development in GI science. This has changed the scope and perspective of GI science from its traditional domain of processing digital spatial data pertaining to maps to processing, visualization of spatial, temporal and spatio-temporal data

Some areas of GI science which have foreseeable research challenges in terms of technology development are:

a) Issues involving modeling of spatial data using soft computing techniques such as AI, NN, fuzzy logic, cellular automata etc.

b) Exploration of the temporal dimension of spatial data to develop analytical tools for spatio-temporal processing and modeling

c) Fusion of spatial data from different sources (sensors and platforms) and agencies with different resolution, scale and information content to generate high quality spatio-temporal information.

d) Development of new algorithms to index spatio-temporal data for querying, processing and visualization in real time and in networked environment; optimization of existing algorithms to reduce the request–response cycle of the user over the network.

References

[1.] Anselin, L.What is special about spatial data? Alternative perspectives on spatial data analysis. Technical Report 89-4. Santa Barbara, CA: National Centre for Geographic Information and Analysis. 1989.

[2.] Buttenfield, B. P. and W. A. Mackaness. Visualization. In *Geographical Information Systems: Principle and Applications* edited by D. J. Maguire, M. F. Goodchild and D. W. Rhind. London: Longman. 1991.

[3.] Chrisman, N. R. and B. Yandell. A model for the variance in area. *Surveying and Mapping* 48: 241–246. 1988.

[4.] Craig, W. J.   URISA's research agenda and the NCGIA. *Journal of the Urban and Regional Information Systems Association* 1: 7–16. 1989.

[5.] Dutton, G. Modeling locational uncertainty via hierarchical tessellation. In *Accuracy of Spatial Databases* edited by M. F. Goodchild and S. Gopal. London: Taylor & Francis. pp. 125–140. 1989.

[6.] Frank, A.U. and D.M. Mark. Language issues for GIS. In *Geographic Information Systems: Principles and Applications* edited by D. J. Maguire, M. F. Goodchild and D. W. Rhind. London: Longman. 1991.

[7.] Franklin, W. R. Cartographic errors symptomatic of under laying algebra problems. *Proceedings, International Symposium on Spatial Data Handling*. Zurich. pp. 190–208. 1984.

[8.] Goodchild, M. F. Keynote address: Progress on the GIS research agenda. *EGIS 91.* Brussels. pp.342–350. 1991.

[9.] Goodchild, M. F. Keynote address: Spatial information science. *Proceedings, Fourth International Symposium on Spatial Data Handling.* Zurich, I, 13–14. 1990.

[10.] Keeper, B. J., J. L. Smith. and T. G.Gregoire.. Simulating manual digitizing error with statistical models. *Proceedings GIS.LIS 88.* Falls Church, VA: American Society of Photogrammetry and Remote Sensing/American Congress on Surveying and Mapping. pp. 475–83. 1988

[11.] Lanter, D. P.   Lineage in GIS: The problem and a solution. Technical Paper 90-6. Santa Barbara, CA: National Centre for Geographic Information and Analysis. 1990.

[12.] Maguire, D. J. A research plan for GIS in the 1990s. *The Association for Geographic Information Yearbook 1990.* London: Taylor & Francis. pp. 267–277. 1990.

[13.] McGranaghan, M. Modeling simultaneous contrast on choropleth maps. Technical Papers, 1991. ACSM/ASPRS/Auto-Carto 10 Annual Convention, Baltimore, MD, March 25–29, 1991, Vol. 2, pp 231–240. 1991.

[14.] Peucker, T. K. and N. R. Chrisman. Cartographic data structures. *American Cartographer* 2: 55–69. 1975.

[15.] Preparata, F. P. and M.I. Shomos. *Computational Geometry: An Introduction.* New York: Springer-Verlag. 1988.

[16.] Samet, H.*The Design and Analysis of Spatial Data Structures.* Reading, MA: Addison-Wesley. 1989.

[17.] Shiryaev, E. E. *Computers and the Representation of Geographic Modeling.* Englewood Cliffs, New Jersey: Prentice Hall. 1987.

[18.] Snyder, John P. *Map Projections: A Working Manual.* Washington, DC: US Government Printing Office. 1987.

[19.] Tomlinson, R. F. Keynote address: Geographical information systems – A new frontier. *Proceedings, International Symposium on Spatial Data Handling.* Zurich. 1: 2–3. 1984.

[20.] Tomlinson, R. F., H. W. Calkins and D. F. Marble. *Computer Handling of Geographical Data.* Paris: UNESCO. 1976.

[21.] White, D. A new method of polygon overlay. *Proceedings, Advanced Study Symposium on Topological Data Structures for Geographic Information Systems.* Harvard. 1977.

Questions

1. Write short notes on evolution of cartography.
2. What are the scientific and technological events that led to the evolution of GI science?
3. What are the contributing disciplines of GI science?
4. What are the distinct technological disciplines and their corresponding tools, techniques and algorithms in GI science?
5. Write short notes on the evolution of GI science.

13

Application Domain of GIS

This book would be incomplete if we did not discuss the application domain of a GIS. Counting and enlisting the applications of an evolving information system like GIS is like counting the feathers of a flying bird. Justice will also not be done to a significant section of readers if the vast range of applications supported by GIS is not discussed. There are numerous applications of GIS and the instances are increasing day by day. This trend is because of the collaborative nature of GIS whereby it can combine data and information from various walks of life and present them with a digital map in the background, making the data easily comprehensible.

Though there are numerous applications, the underlying principles governing most of these applications are similar. All these applications can be clustered and classified into different sub and sub-sub-sets depending upon the similarity of usage and principles of GIS they use. Nevertheless each application of GIS can separately be a project experience, and be put down as a chapter in itself. To make the study more systematic, instead of explaining the applications individually, the generic concepts behind a large set of similar applications have been categorized and their functioning discussed. I have tried to list the generic areas of applications of GIS and categorize them. Some of the important applications are presented with illustrations.

Block diagrams, sequence diagrams, collaboration diagrams and process flow diagrams are well-defined mechanisms to describe a process. They clearly bring out the involvement of sub-processes, and user interactions with various devices of an overall system. Hence to explain the application of GIS as a system I have made use of these concepts in the later part of this chapter.

13.1 Classification of Application Domain

The application domain of GIS can be classified into two broad categories: (a) defense/military application and (b) civil/commercial application. These broad categories can be further classified into sub-and sub-sub-categories, which are highly cohesive in nature. The sub-category application area is depicted in the hierarchical tree diagram (Fig.10.1). Although the examples are not exhaustive, the generic candidates in the application areas have been enlisted.

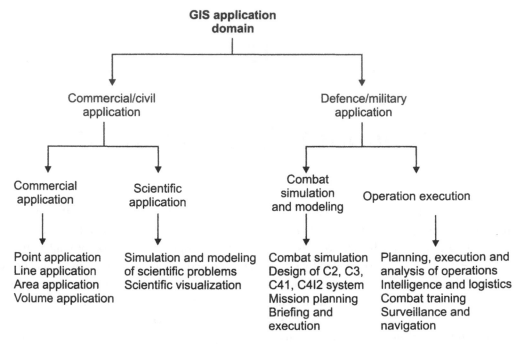

Fig. 13.1 Application categories of GIS

The specific sub-application areas under these major application categories are given below

<u>Simulation and modeling</u>
Combat simulation
Flooding simulation.
Earthquake simulation
Fly through simulation
Land movement and plate tectonics

<u>Scientific visualization</u>
Spatial DSS
Modeling and simulation using spatial data
Scientific visualization of natural hazard
Simulation of nuclear effect of nuclear warhead

<u>Commercial application</u>
Location services
Fleet management
Forest survey management
Wild life survey management
Geological exploration and mining application
Mine management → Mineral deposit calculation
Hydrology application → Water resource planning
Fresh water management

<u>Resource management</u>
Water resource management
Land usage pattern, urban land distribution pattern
Rural land for agriculture surface
Transportation/road and bridge planning
Mining → Mineral exploration, excavation and mining

Telecomm application → Radio line of sight for communication
Finding appropriate location of transmitter and receiver for terrestrial communication network
Educational applications → preparation of educational maps
Tourist guide map, thematic maps and elevation models
Spatial data fusion (remotely sensed data, surveyed data and online data)

13.2 Military Applications of GIS

During a battle, spatial data is of crucial importance to the military commander for operation planning. That is why the Ministry of Defense (MOD) in any country has a special organization which looks after collection, collation, synthesis and ground truthing of spatio-temporal data. These are performed through various sources and agencies e.g. aerial photography, ground survey etc. Saptial data acquisition is performed periodically to get updated terrain information. GIS helps in visualizing, analyzing, filtering and interpreting spatial data for effective decision-making. Visualizing raw tabular data within a spatial framework has many benefits. Therefore, digital mapping and geographic information system (GIS) occupy center stage in activities as diverse as battlefield simulation, mission briefing, communications planning, and logistics management and command control.

Computer-based geographical information systems can provide automated assistance to military forces terrain analysis function. However these software systems and utilities have limitations as they are not full-featured GIS. The greatest limitation is the dependence on the user's ingenuity and the data which is used. These systems have the capability to receive, re-format, create, store, retrieve, update, manipulate and condense digital terrain data to produce terrain analysis products such as modified combined obstacle overlays, hydrology overlays, slope maps, on and off road mobility maps, line-of -sight plots, concealment maps and possible problems associated with lines of communication.

The uses for GIS will continue to evolve as technology advances and the costs decreases. The full potential of some GIS applications in military forces has already been discovered, however the future of GIS applications in the military will be determined by how military units accept GIS and utilize it in the most efficient way possible. Present warfare involves operations with combined forces and an integrated approach for evaluating battle area for mobilizing logistics, moving various forces and setting communication networks. Effective operations in real-time scenarios are necessary prerequisites for successful operations.

GIS technology helps armed forces, as information is readily available to various levels of officers involved in operations. The advent of remote sensing technology has proved to be a great boon to intelligence units in defense forces by acquiring data on enemy activities from so-called 'eyes in the sky'. Spy satellites constantly acquire high-resolution satellite data, even during times of peace, to monitor the development in acquisition of modern warfare gadgets by the enemy forces. There is no privacy as far as these satellites are concerned. In fact, developed countries have been extensively using remote sensing techniques to monitor enemy activities in establishing nuclear installations. This information is then brought to the notice of international agencies coordinating the prevention of use of nuclear energy for destructive purposes.

The use of remote sensing data combined with ground information would provide a common platform for analyzing the ground situation in times of war. The induction of satellites providing high-resolution images in the present era enhances the ability to make maps accurately and to make the latest information available to the forces.

13.2.1 Usage of GIS in the Army

The concept of command, control, communication and coordination in military operations is largely dependent on the availability of accurate information in order to arrive at quick decisions for operational orders. Geographic information systems (GIS) play a pivotal role in military operations, as they are essentially spatial in nature. The use of GIS applications in defense forces has revolutionized the way in which these forces operate and function. Military forces use GIS in a variety of applications which include cartography, intelligence, battlefield management, terrain analysis, remote sensing, logistic management etc. GIS is also being used in counter-insurgency operations for monitoring of possible terrorist activity. An overview of the use of GIS in military applications in land-, sea- and air-based operations is attempted in subsequent sessions. GIS has evolved as a mechanism to aid the battle manager coordinate, command, and control operations and communicate the unified view of the battle to the field forces.

How important is spatial information to the field commander at command headquarters to take an appropriate decision for military operations can be understood by the examples provided in the subsequent sections. GIS can be said to be the heart of many military systems. GIS is especially being used in tactical C2, C3, C3I2 and C4I2SR systems which are used as virtual force multipliers (Here the 4 Cs stand for command, control, communication and coordination, the 2 Is for information and interoperability, the S for space and the R for reconnaissance. So a C4I2 system is a command, control, communication, coordination, information, interoperable system). GIS being a collaborative information processing system, the importance of interoperability, which is a very important aspect in military scenarios, cannot be over-emphasized. GIS acts as binding software for heterogeneous armament and software systems inducted in military operations. GIS also plays a vital role in achieving interoperability of information among heterogeneous combat groups. Specific ways of GIS application in different forces is discussed in subsequent sessions.

13.2.2 Use of GIS in Navy and Maritime Operations

Protecting and defending the maritime border and economic zone of a country is as important as protecting its land border. Some countries are surrounded by sea; thus not needing to share a land border with any other country but having an extensive seacoast. Hence it becomes important to update and manage the spatial information regarding the maritime border.

Countries having sea borders transact over 98% of the nation's cargo through waterborne transportation. Therefore, the navy or marine force plays an important role in defense. Some important uses of geo-spatial information by any marine force are:

1. Monitoring and defending the economic and maritime border
2. Preparation and usage of electronic navigation charts, nautical charts, mariner's charts and hydrographic charts.
3. Coast zone mapping and change detection
4. Island and seabed mapping
5. Naval combat simulation
6. Monitoring of sea hazards such as tsunamis, tidal waves due to cyclone.

Naval Operations

In sea, vessels depend largely on indirect methods such as pole star, magnetic campus, islands and lighthouse indicators etc. to navigate when there is no means of establishing their position with visual aids. Global positioning systems (GPS) provide a more accurate means of determining position at sea. Echo sounders measure the depth of the water below the vessel. Naval vessels operate at sea using several electronic gadgets for operations. Recent technological advances have provided the means to assess the unknown to greater accuracy. In the oceans, complex natural features such as currents, wave conditions, sea surface temperatures and tides may prove deterrent at times to naval operations. Using these natural features to obtain a clear understanding of the complex ocean dynamics is an essential element for successful naval operations. All these natural phenomena can be viewed after superimposing them against the backdrop of a sea map for a collaborative picture.

The recent induction of the electronic chart display and information system (ECDIS), on the bridge of vessels, helps the navigator to navigate the ship safely in all weather conditions. The electronic navigation chart (ENC) is a replacement of the conventional paper chart, which is used as tool for navigation. It provides inputs for detailed information about depth, hazards and navigational aids within the area. This supported by visual and audio alarms of ECDIS provide the navigator sufficient means to navigate the vessel safely. The display is used to provide selective spatial or textual information to the navigator for safe passage. Thus, ENC is the database for GIS operations and ECDIS is the real-time GIS application in the marine environment.

In addition, ECDIS can be used for other naval operations such as anti-submarine or beach landing of armed forces in military operations by adding supplementary layers of information related to oceanographic and meteorological conditions. However, at present in the marine environment, the use of ECDIS is limited to navigation and most countries are switching to ENCs of their maritime zone, the production of which itself is a very costly and time-consuming operation. A beginning has been made to create these datasets. It is expected that in another five years time the attention would be towards the use of ENC with additional layers for both military and scientific applications.

13.2.3 Use of GIS in the Air Force

Air operations in a battle environment require similar inputs as land operations, but also precise height information for targeting. These include detailed information about the target location, proximity of civilian areas, terrain evaluation and meteorological conditions besides navigational data. Virtual reality concepts are of great help in fighter and bombing aircrafts for effective air strike operations.

Military leaders heavily depend on GIS and GPS (global positioning systems) to make tactical decisions such as guiding troops, supplies/equipment and ships, informing them of possible threats, problems with terrain which they will encounter and also to direct their attention to specific areas of interests. For example, data is relayed to the attack aircraft giving the pilot needed information such as the location and identification of the target, plus possible hot spots in which they may encounter an attack on themselves. These pilots also receive data relating to meteorological information, which enhances visibility, and pre-warns them of possible change, which may occur during an aerial activity. Some specific uses of GIS in the air force are:

1. Demarcation of fly zones and no fly zones
2. Target tracking and visualization
3. Air operation planning and coordination
4. Air traffic control and navigation
5. Fly through simulation
6. As a GUI to the cockpit of a flying trainer

13.3 GIS in Navigation

Use of GIS for navigation is one of the most widely used applications of GIS. Various navigation charts, which are the fundamental tools available to the mariners, are available in various forms and formats as described below. Different navigational charts are prepared using GIS as the backbone software with other electronic display and sensors such as

1. Nautical chart (NC).
2. Electronic navigational chart (ENC)
3. Raster nautical chart (RNC)
4. Hydrographic charts
5. Aeronautical charts

13.3.1 Electronic Navigation Chart (ENC)

ENC is used in vessel traffic systems (VTS) to monitor ship movements in rivers, harbors, and bays. GIS provides spatial data for background display and acts as a GUI (graphical user interface) of ENC which is used for marine navigation, route planning etc. ENC uses vector spatial data constituting of many thematic categories such as sea bottom contours, navigational obstacles, sea current maps etc. Users can choose one or a combination of different thematic categories to display and compose the background map.

13.3.2 Nautical Chart

A nautical chart is a graphic portrayal of the marine environment showing the nature and form of the coast, the general configuration of the sea bottom including water depths, locations of dangers to navigation, locations and characteristics of man-made aids to navigation and other features useful to the mariner. The nautical chart is essential for safe navigation. In conjunction with supplemental navigational aids, it is used by the mariner to layout courses and navigate ships by the shortest and most economically safe route. The importance of nautical charts is evident with over 98% of the world cargo being carried by waterborne transportation.

A raster nautical chart (RNC) is a geo-referenced, digital image (Fig. 13.2) of a paper chart, which can be used for navigation in sea. Today, there are many GIS applications which display the nautical chart along with its S-57 vector map and GPS to give online navigational aids.

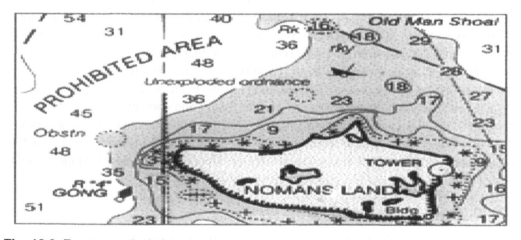

Fig. 13.2 Raster nautical chart

13.3.3 Benefits of Using ENC

Many marine mishaps are due to human error. Incorporating digital chart data with continuous GPS signals for automated vessel positioning enhances safety of navigation. Users can selectively display only the information desired while the computer can continue to process all the information for safety of navigation. Vector chart data with appropriate GIS software applications can provide the mariner with advance electronic warnings of unforeseen dangers. ENC are easy to maintain by upgrading current spatial data and software with new features.

13.4 Specific Applications of GIS in Defense

13. 4.1 Defense Estate Management

The value of this GIS technology as an administrative support is significant, particularly if the Ministry of Defense (MOD) of a nation holds large chunks of property across the nation. Effective management of the defense estates requires significant effort. In this background, the MOD with its three wings of armed forces is served by mapping all its spatial assets in both digital and analogue forms. GIS helps in storing, updating and retrieving all these data. Use of GIS in the management of military bases facilitates maintenance and the tackling of all stores, which may be found on the base. GIS allows military land and facilities managers to reduce base operation and maintenance costs, improve mission effectiveness, provide rapid modeling capabilities for analyzing alternative strategies. It helps to improve communication and store institutional knowledge for post-battle analysis and training.

13.4.2 Terrain Evaluation

In land-based military operations, it is important for military field commanders to know terrain conditions and elevations for maneuvering armor carriers, tanks and for the use of various weapons. In addition, they need information on vegetation cover, road networks, and communication lines with pin-point accuracy for optimizing resource utilization. A detailed land map with information on the land use, terrain model and proximity of habitat are essential for military operations. All these details must be available to the field commanders on a datum to match with the equipment he uses for position fixing and communication in his area of operation. Any discrepancy in these inputs may endanger the operation. Target assessment can only be done if the inputs properly match the system used for firing the weapon. Magnetic variation and gravitational information are also required for sensitive military operations (Birkin et al. 1996).

13.4.3 Viewing Spatial Data

Most potential users of GIS are viewers. There are personnel right from field commanders to command staff. They need access to geographic pictures, maps or photographs to help and assess a situation to carry out planned operations. Earlier GIS packages were proprietary in nature and restricted the use of data within its confined specifications. A comprehensive database in multi-type data integration background needs an open GIS approach. An open GIS approach allows a choice of the most appropriate product for individual users and at the same time supports command requirements to specify an authorized map for operational reasons.

13.4.4 Weather Information

Weather plays a dominant role in the battlefield. Real-time weather information is essential for field commanders on land, in sea or in air for successful completion of the task. At times, weather may play a crucial role in the success or failure of an operation. Every battlefield commander would like to know information regarding cloud coverage, wind conditions, visibility, temperature parameters and other related inputs. Also required for successful operation is the almanac data of the place of operation such as time of sun rise, sun set, moon rise, moon set, quarter of moon etc.

13.4.5 Positional / location Information

One of the most important functions of GIS along with satellite imagery is to understand and interpret terrain. This function has a major role in determining how troops can be deployed in the quickest and most effective way. Understanding the landscape is especially useful because a military leader can determine strategic positions, such as ideal locations for scouting parties, best line of sight/fire and also the ability to hide troops and equipment.

13.4.6 Logistics Management

GIS plays an important role in military logistics because it helps in moving supplies, equipment, and troops where they are needed at the right time and place. By using GIS for determining routes for convoys, forces are able to determine alternative routes if mishaps or traffic jams occur on the most direct route. By using both GPS and GIS, certain sensitive articles such as nuclear warheads can be tracked every step of their shipment and also kept away from hot spots, populated areas and other shipments.

13.4.7 Mapping Techniques

From the above it is evident that the military needs different maps for different purposes within its operational command and each requirement is to cater for a specific purpose. The digital base in GIS environment facilitates the creation of different types of maps to meet specific user needs without clustering with unwanted details. This facilitates the viewing of spatial information on a need-to-know basis either at command headquarters or in the field area. The battle commanders can evaluate thematic information for analyzing the real-time scenario by manipulating the information available at their disposal.

13.4.8 Common Horizontal Datum

It is necessary that the spatial data used by military units reside within a framework of single datum for coordinating joint service operations. In the present scenario of military operations, there is a bottleneck in this aspect because 'maritime operators such as navy and marine forces use a vertical datum based on the high water mark; land operators such as military use a vertical datum based on mean sea level, while air operators like air force use obstruction heights above ground level. Amphibious operators have to take note of this fact and normalize the use of datum so that the computation of location of battle elements is the same and is accurate.' Different projections and datum are used by different agencies to achieve their objectives as given in Section 10.7.

The problem becomes more complex when multinational forces are deployed. This was evident in many joint operations where there was a difference in target position computed by European Datum and WGS84 by a few hundred meters. Hence weapon system and forces using different datum need to synchronize themselves before the start of a joint operation. In India, the defense forces use Everest Datum and GPS receivers use WGS84 datum. Hence use of GPS receivers in the field may pose problem unless the datum shift is correctly established. There should be interoperability between the three wings of the armed forces to use a common reference datum in their activities while mapping for effective conduct of joint military operations.

All this boils down to the fact that a common datum is necessary. Slowly WGS84 is emerging as a common datum for all such operations. The technological advances in position fixing using satellites are based on WGS84 and most civilian applications need to be shifted to this datum in course of time. Military applications are no exception. However, this is a gigantic task and to achieve a common datum across the world needs money and expertise. Most countries may not have the resources in terms of funding and technology to handle this change.

13.5 Commercial Application of GIS

There are countless instances of commercial exploitation of GIS and their usage in civil applications such as disaster management, emergency response etc. Enlisting them and categorizing them convincingly is difficult because of the similarity of the usage and new emerging fields of applications. But roughly, use of GIS in civil applications can be categorized into three main topological categories involving point, linear and area applications. All the applications of GIS can fit into any one or combination of the categories depicted in Fig.13.3.

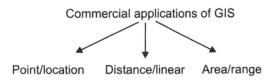

Fig. 13.3 Different categories of commercial applications

13.5.1 Point/location Applications

A major set of commercial applications is about locating a suitable resource among spatially distributed locations e.g. hospitals, hotels, shopping malls etc. Choosing a suitable location out of many is challenging and computationally complex. GIS solves these problems by applications of spatial queries and filters. The spatial queries make use of computational geometric algorithms such as a point inside a triangle, polygon, circle etc. In solving these queries, GIS acts as a spatial decision support system (Birkin et al. 1996) by suggesting ideal locations to establish resources for operational purposes. Some of the important applications in this category are:

1. Suitable location for sighting communication transmitters and receivers.
2. Suitable location to establish a facility e.g. fire station, medical facility so that it can be utilized by a vast part of the locality or population.
3. Suitable location for breaching a water body so as to get the desired effect of flooding.

13.5.2 Distance/linear Applications

Use of GIS for applications to find suitable routes, length and distance between a source and destination has lots of commercial implications. GIS is used for computation of optimum path in a communication network. Sometimes, all possible routes, shortest routes, optimum routes, alternate route computation finds much business application in transportation. GIS solves these problems by making use of graph theory algorithms such as Dijkstra's algorithm and Floyd's algorithm. Computing distance has many connotations in GIS and it is being used to compute the cumulative distance, crow fly distance and shortest distance between two communication points. Some of the major applications involving linear computation in GIS are

1. Navigation and way point and routing of communication vehicles
2. Comuting shortest, optimum and alternate routes in a communication network.
3. Land, sea and air traffic and fleet management

13.5.3 Area/range Applications

Applications involving area computation in GIS has many applications as spatial decision support systems. GIS helps in solving many operation research problems involving spatial distribution and suitable areas of many domains. Typical questions answered are 'find the suitable area for construction of colony', 'find the area affected by a particular disease', 'find the area served by a particular public utility or facility are being solved using GIS'. GIS makes use of buffer query and filters designed using computational geometric algorithms to solve such problems. Some prime applications involving area analysis are

1. Suitable area for staging of a moving convoy
2. Identifying suitable area for construction of a stadium
3. Identifying suitable area for construction of public utility
4. Identifying suitable area for growing of a specific type of crop etc.

Besides the above applications, GIS is being extensively used in applications such as spatial data conversion, forestry and wild life conservation, hydrology and water conservation, geological exploration and mining, study of land use patterns, remote sensing, hazard monitoring and emergency response, environmental conservation and pollution monitoring, city planning, map publishing and GIS education etc.

13.6 Management Application Areas of GIS

GIS are now used extensively in government, business, and research for a wide range of applications including environmental resource analysis, land use planning, locational analysis, tax appraisal, utility and infrastructure planning, real estate analysis, marketing and demographic analysis, habitat studies and archaeological analysis.

One of the first major areas of application of GIS was in *natural resources management*, including management of

- wildlife habitat
- wild and scenic rivers
- recreation resources
- floodplains
- wetlands
- agricultural lands
- aquifers
- forests

One of the largest areas of application has been in *facilities management*. Uses for GIS in this area includes

- locating underground pipes and cables
- balancing loads in electrical networks
- planning facility maintenance
- tracking energy use

Local, state, and federal governments have found GIS particularly useful in *land management*. GIS has been commonly applied in areas like

- zoning and sub-division planning
- land acquisition
- environmental impact policy
- water quality management
- maintenance of ownership

More recent and innovative uses of GIS have used information based on *street-networks*. GIS has been found to be particularly useful in

- address matching
- location analysis or site selection
- development of evacuation plans
- analysis for shortest path, optimum path, alternate path in a communication networks

References

[1.] Abel, D. J., P. J. Kilby and J. R. Davis. The system integration problem. *International Journal of Geographical Information Systems* 8:1–2. 1994.

[2.] Ackermann, F. Experimental investigation into the accuracy of contouring from DTM. *Photogrammertic Engineering and Remote Sensing* 44:1537–8. 1978.

[3.] Andes, N. and J. E Davis. Linking public health data using geographical information system techniques: Alaskan community characteristics and infant mortality. *Statistics in Medicine* 14: 481–90. 1995.

[4.] Anselin, L. and A. Getis. Spatial statistical analysis and geographic information systems. In *Geographic Information Systems, Spatial Modeling, and Policy Evaluation* edited by M. M. Fischer and P. Nijkamp. 35–49. Berlin: Springer. 1993.

[5.] Batty, M. Using GIS for visual simulation modeling. *GIS World* 7(10): 46–8. 1994.

[6.] Birkin, M., M. Clarke, G. Clarke and A. G. Wilson. *Intelligent GIS: Location Decisions and Strategic Planning.* Cambridge (UK): GeoInformation International. 1996

[7.] Brimicombe, Allan. *GIS, Environmental Modeling and Engineering.* George Green Library. 2003.

[8.] Densham, P. J. Spatial decision support systems. In *Geographical Information Systems: Principles & Applications* edited by D. J. Maguire, M. F. Goodchield and D. W. Rhind. vol. 1:403–12. Harlow: Longman/New York: John Wiley & Sons Inc., 1991.

[9.] Easa, Said and Yupo Chan (eds.). *Urban Planning and Development Applications of GIS.* Reston, Va: American Society of Civil Engineers. George Green Library TD160 URB, 2000.

[10.] Enache, Miracea. Integrating GIS with DSS: A research agenda. *URISA* 154–166. http://wwwsgi.ursus.maine.edu/gisweb/spatdb/urisa/ur94015.html, 1994.

[11.] Federa, K. and M. Kubat. Hybrid geographical information systems. *EARSel Advances in Remote Sensing* 1:89–100. 1992.

[12.] Goodchild, M. F. A spatial analytical perspective on geographical information systems. *International Journal of Geographical Information Systems* 1:321–34. 1987.

[13.] Goodchild, M. F., R. P. Haining and S. Wise. Integrating GIS and spatial data analysis: Problems and possibilities. *International Journal of Geographical Information Systems* 6:407–23. 1992.

[14.] Keenan, Peter. Using a GIS as a DSS generator. *Working Paper Management Information Systems* 95–9. Michael Smurfit Graduate School of Business, University College Dublin. http://mis.ucd.ie/staff/pkeenan/gis_as_a_dss.html,1997.

[15.] Lyon, John G. *GIS for Water Resources and Watershed Management.* Hallwood Library. 2000.

[16.] Maguire, D. J. Implementing spatial analysis and GIS applications for business and service planning. In *GIS for Business and Service Planning* edited by P. A. Longley and G. Clarke. 171–91. Cambridge (UK): GeoInformation International. 1995.

[17.] Skidmore, A. *Environmental Modeling with GIS and Remote Sensing.* Taylor & Francis. Hallward Library. 2001.

[18.] Steyaert, L. T. and M. F. Goodchild. Integrating geographic information systems and environmental simulation models. In *Environmental Information Management and Analysis: Ecosystem to Global Scales* edited by W. K. Michener, J. W. Brunt and S.G. Stafford. 333–57. London: Tayler and Francis. 1994

[19.] Warren, I. R. and H.K. Bach. MIKE 21: A modeling system for estuaries, coastal water and seas. *Environmental Software* 7:229–40. 1992.

Questions

1. What is the application domain of an information system in general and GIS in particular?
2. What are the different generic categories of GIS applications?

3. Describe the uses of GIS in the civil industry with examples.
4. Describe the uses of GIS in the defense industry with particular reference to land, air and water forces?
5. Describe the applications of GIS in the field of navigation.
6. What are NC, ENC, and RNC?

14

Spatial Decision Support System (SDSS)

GIS can correlate space, time and attribute data—its input domain includes objects in space, earth and sea. Therefore GIS has emerged as a collaborator of data pertaining to objects which are geo-referenced over the entire globe. Any problem involving spatial data with or without temporal variance can be studied using GIS. GIS can also discover patterns from a vast amount of spatio-temporal data and provide information regarding variation of spatial patterns. This makes GIS a good research platform for spatial data mining and spatial pattern discovery. In a sense, GIS helps in discovering knowledge regarding spatial variance of objects.

The concept of decision support evolved from the theoretical studies of organizational decision-making done at the Carnegie Institute of Technology during the late 1950s. DSS became an area of research on its own in the mid-1970s, before gaining intensity as part of computer science and information systems during the 1980s (Power 2003).

Since GIS is a must for any problem involving spatial data, its characterization, analysis, modeling, visualization and management, it has emerged as a basis platform for management and decision support for problems involving spatial data. This chapter explores the emergence of GIS as a spatial decision support system (SDSS). It starts with the process of decision making and the sequence of steps involved in decision making. The definition of DSS and the extension of the concept to SDSS follow with a comparison of DSS vs. SDSS. Various components of SDSS are discussed with a block diagram to make the concept clear. Applications of SDSS and its benefits are discussed. Finally various examples of SDSS implemented by the author are given.

14.1 Decision Support System (DSS)

The majority of information technology applications exploit non-spatial data for generation of information. About 90% of them are engaged in collection, collation, storage of data in database and retrieval through query. A typical user interacts with such system through a

GUI, which gives him the option to interact and retrieve data through a set of pre-designated queries. The response is in the form of a textual report tabulating the data in a matrix or a graph correlating different sets of data. In a sense, these systems provide information by structurally presenting the data to a user's query. The block diagram of a typical information system is depicted in Fig.14.1. These systems are often referred to as management information systems (MIS). In these types of systems, comprehending and deriving meaningful information from the reports is left to the user's discretion.

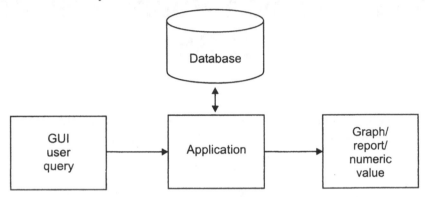

Fig. 14.1 Components of a management information system

14.1.1 What is a DSS?

A DSS is different from a normal information system—it works at higher levels of information processing. DSS applies rules to generate contextual data from the database. It further applies semantic rules and decision rules to the contextual data to generate decision options or alternatives. In a sense, it solves the problems posed by a user and gives decision alternatives. Generally, the problems are highly domain specific such as medical diagnosis, resource optimization, suitable weapon system usage etc. The building blocks of a DSS constitute a

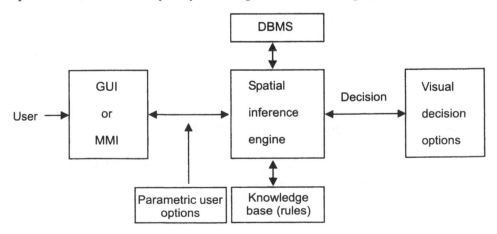

Fig. 14.2 Components of a decision support system

GUI, which receives user inputs and formulates the problem. The problem can be modeled as a set of queries fired to the database in a cascaded manner. The queries can be logical or parametric in nature. The inference engine fires the rules to the contextual data to derive decision outputs. A schematic block diagram of a DSS is depicted in Fig.14.2.The decision is conveyed to the user through a visual display or through a MMI (man–machine interface).

14.1.2 Stages of Data Transformation in DSS

The various stages involved in transformation of data in a DSS are depicted in Fig.14.3. The data is transformed to a decision through various stages known as DIKD (data–information–knowledge–decision) chain. The arrows in the figure represent the means of transformation of the data from one state to another.

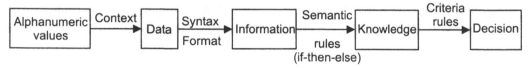

Fig. 14.3 DIKD chain of information transformation in DSS

14.1.3 Types of DSS

There are many approaches to decision-making because of the wide range of application domain in which decisions are made. That is why; *decision support system* (DSS) can take many different forms. In general, we can say that a DSS is a computerized system for helping to make decisions. A decision is a choice between alternatives based on estimates of the values of those alternatives. Supporting a decision means helping the user to gather intelligence, generate alternatives and make choices. It can also be defined as a process which involves supporting the estimation, the evaluation and/or the comparison of alternatives. Hence there is no single comprehensive definition of DSS. Some important definitions of DSS are given below.

DSS are a class of computer-based information systems including knowledge-based systems that support decision-making activities. Alternatively DSS are 'interactive computer-based systems that help decision makers utilize data and models to solve unstructured problems.'

More specifically, a DSS is a software application which assists the user in collecting, collating, analyzing, modeling and filtering solutions pertaining to the problem and which assists the decision maker with alternatives. It utilizes data, provides an easy-to-use interface, and allows the decision maker to add his own insights.

DSS can also be defined as a model-based set of procedures for processing data and judgments to assist a manager in his decision-making. DSS couples the intellectual resources of individuals with the capabilities of the computer to improve the quality of decisions.

In a nutshell, DSS are a computer-based support system for management decision makers who dealing with semi-structured problems.

Decision-making is highly specific to the domain or class of applications. Hence, DSS are application-specific software providing solutions to solve problems in a particular domain. For example, a medical diagnostic DSS cannot handle trade-related issues of the stock market. DSS are therefore engineered depending upon the application domain and approach adopted to solve the problem, type of input data processed etc. Using the mode of assistance as the criterion, DSS are categorized (Power 1997) as communication-driven DSS, data-driven DSS, document-driven DSS, knowledge-driven DSS, and model-driven DSS.

1. Communication-driven DSS

Communication-driven DSS supports more than one person working on a shared task. These systems are necessarily collaborative processing systems where the decision collaborators are interconnected through an enterprise-wide DSS in a network. The intermediate decisions and alternatives are refined and improved through a hierarchy of decision makers belonging to the same organization. Examples of communication-driven DSS include integrated tools like Microsoft's NetMeeting or Groove.

2. Data-driven DSS

Data-driven or data-oriented DSS emphasizes access to and manipulation of a time series data of an application or organization. It sometimes intakes external data and parameters for the decision-making process.

3. Document-driven DSS

This type of DSS manages, retrieves and manipulates unstructured information in a variety of electronic formats.

4. Knowledge-driven DSS

Knowledge-driven DSS provides specialized problem-solving expertise stored as facts, rules, procedures, or similar structures.

5. Model-driven DSS

This type of DSS emphasizes access and manipulation of a statistical, financial, optimization, or simulation model. Model-driven DSS use data and parameters provided by users to assist decision makers in analyzing a situation; they are not necessarily data-intensive. A mathematical or parametric model is developed from a part of the data. It is then validated using the rest of the data. The proven model is then used to derive solutions to the problem posed by the user.

Using scope as the criterion, DSS can be categorized as an enterprise-wide DSS and desktop DSS. An *enterprise-wide DSS* collaborates simultaneously with a large number of stake holders through a LAN or WAN. It is linked to large data warehouses and serves many managers in the organization to collaborate in the decision-making process. A *desktop, single-user DSS* is a small system that runs on an individual manager's PC.

14.2 Definition of SDSS

A spatial decision support system (SDSS) is a DSS, which processes spatial and temporal data in addition to non-spatial data traditionally being processed by a DSS. It generates information and decision options pertaining to problems involving spatio-temporal information.

It differs from a normal DSS as it generates the decision outputs in the form of visual simulation besides generating textual and graphical reports. It gives many alternatives to the problem under consideration. One can make use of its cognitive decision-making capability in analyzing reports. Thus SDSS can give different views and alternatives to the results i.e. numerically, visually, textually, tabular manner or in the form of a graph.

Loosely, SDSS can be defined as an interactive, computer-based system designed to support a user or group of users to achieve a higher effectiveness of decision-making while solving semi-structured spatial problems. It can assist spatial planners make useful decisions regarding land use. For example, when choosing the site of a new airport, many contrasting criteria, such as noise pollution vs. employment prospects or the knock-on effect on other transportation links, makes the decision difficult. A system which models decisions could help identify the most effective decision path.

SDSS can be alternatively defined as a special class of DSS with mechanisms for solving problems involving spatial data. It can analyze and visualize decisions which are unique to spatial and geographical data through spatial analysis tools, geo-statistics and spatial interpolation techniques.

GIS is the kernel of SDSS as it is capable of processing spatio-temporal data. Administrators, engineers, designers and users of GIS systems in general encounter problems which are spatial in nature. In practice these problems are ill-structured or semi-structured with lots of abstraction and degenerate data. A GI problem can be characterized by the following dimensions:

- It involves spatial dimension
- It is ill-posed or semi-structured
- There are multiple possible outcomes which are equally probable and conflicting in nature.

GIS emerges as a strong contender to solve problems which are ill- or semi-structured, and involves complex spatial decision-making because it can capture the important dimensions of spatial problems using mathematical models and hybrid mathematical formulations. GIS can provide a decision process to evaluate various alternatives and reconcile conflicting solutions. It can provide the decision maker with a flexible problem-solving environment to increase the level of understanding of the problem and evaluate alternate solutions. One important aspect of GIS is that it is a platform for thematic analysis of spatial data by which it can define a relationship between spatial components through overlays and maps. This is where GIS acts as a SDSS (spatial decision support system).

GIS alone is *not* a decision-making tool kit because making (good) decisions require
- knowledge and foresight
- insight and intelligence

- expertise, i.e. rational choice between alternatives (especially where conflicts are present).

GIS does not provide the above, but it can fulfil an important role in decision making by providing decision support through its spatial data input capabilities, storage of complex structures and analytical techniques unique to spatial data. GIS can provide the much needed

- geographic information analysis capabilities
- cartographic computations and output
- spatial analytical modeling
- Spatial data display capability

14.3 Inputs of SDSS

An SDSS intakes spatial and non-spatial data from different sources and agencies and collates them to a common frame of reference i.e. common coordinate system and datum. Figure 14.4 depicts a set of spatial data obtained from different sources, which can be processed by SDSS.

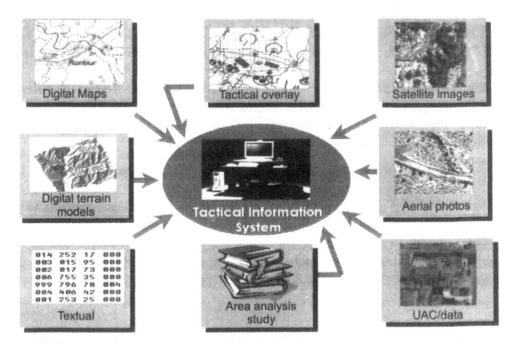

Fig. 14.4 Spatial inputs to a SDSS

An SDSS can process spatial data specified in the input domain of the GIS process (Chapter 2, Table 2.1). Besides this it can also intake user-defined parameters and data to formulate the problem for which decision solutions are sought.

14.4 Architecture of SDSS

The architecture of SDSS decides the availability of the system resources and services to the decision maker. It also decides the scalability, throughput, reliability and security of geo-spatial data, information and decisions. It gives a clear picture of the various sub-systems and their roles in the SDSS. Therefore GIS architecture plays an important role in the choice of GIS. In fact the architecture of SDSS holistically considers the hardware, software and information architecture of the system as a whole and the sub-systems as parts. SDSS designers must understand every aspect of the architecture before implementing a system. The architecture of an SDSS is depicted in Fig. 14.5.

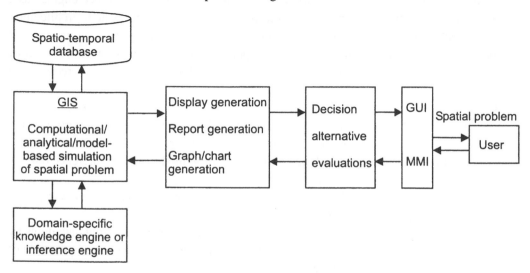

Fig. 14.5 Architecture of a spatial decision support system

14.4.1 Spatio-Temporal Database

The spatial database is the repository of the application domain's spatial data. These are the internal and external data that will contribute to the decision-making process. The data are time stamped i.e. the time of its collection or capture is declared along with the data. A typical SDSS contains the processed input domain (Table 2.1), the domain-specific spatial data and attribute data such as census, deployment, vehicle data etc. In most cases, this data is more extensive and complex than traditional relational models. Often the spatial database must differentiate between the levels of inputs and check the quality of the data.

14.4.2 Model Base

This module of an SDSS contains a set of algorithms that makes decisions based on the information in the database and parameters inputted by the user. It is highly domain-specific because the algorithms and models of a particular domain are not applicable to any other domain. For example, the model predicting demographic trends will not be applicable for predicting the growth pattern of an urban area or for that matter predicting the trend of the stock market. The model base of an SDSS makes use of soft computing tools such as neural networks (NN), artificial intelligence (AI), fuzzy logic and cellular automata (CA) for modeling spatio-temporal data. These tools take a part of the input data for training and stabilizing the model. This information is then summarized and displayed as tables or graphs.

14.4.3 Knowledge/Inference Engine

The knowledge base or inference engine is the heart of the SDSS. It constitutes a set of decision rules specific to the application domain. The separation of inference engines as a distinct software component originates from the typical production system architecture. This architecture relies on a spatial data store, or working memory, serving as a global database of symbols representing the spatio-temporal assertions of the problem. The set of rules which constitute the program of the inference engine, and which it is required to execute (executing rules is also referred to as firing rules), is stored in a rule memory or production memory. The inference engine determines which rules are relevant for a given spatial data set stored in the spatial database of the SDSS and even chooses the portion of the spatial data on which the rules will be applied. The software control mechanism which selects the set of rules is often called the conflict resolution module of the inference engine.

An inference engine has three main elements:

1. *An interpreter*   The interpreter chooses the set of spatial data items by applying the corresponding base rules.
2. *A scheduler*   The scheduler maintains control over the spatial data items by estimating the effects of applying inference rules in the light of item priorities.
3. *A consistency enforcer* The consistency enforcer attempts to maintain a consistent representation of the computed solution.

The inference engine can also be described as a finite state machine with a cycle consisting of three action states: match rules, select rules, and execute rules. At the first state, *match rules*, the inference engine finds all the rules that are satisfied by the current contents of the data store. The rules are in the typical *condition–action* form; this means the condition is tested against the working memory. If a rule is matched to some data, the rule is referred to as a candidate rule or a *conflict set* for execution. The same rule may appear several times in the conflict set if it matches different subsets of data items. A pair consisting of a rule and a subset of matching data items is called an *instantiation* of the rule. The select rules state selects the appropriate rules for the problem under consideration and sequences them for execution. The execute rule state applies the rule to the database and obtains the candidate set of output.

14.4.4 Graphical User Interface (GUI)

GUI of SDSS plays a crucial role in making it usable and popular. A GUI takes in the command and domain-specific parameters of the DSS. It is an important part of the SDSS because it is through the GUI that the initiation of a problem is carried out; the user interacts and interprets intermediate and final results through a GUI by means of visual and graphical displays.

The GUI has a challenging role to perform. Often the user who interacts with the system is inexperienced. A crucial segment of users of SDSS are physically challenged such as deaf, dumb and blind. Certain sophisticated users like a pilot in the cockpit of an airplane cannot spare his hands during a mission to give commands or interact with the GUI of SDSS. Hence, nowadays sophisticated sensor-based input/output components are attached to the GUI, making it a true MMI (man–machine–interface), for example, voice command systems which intake the commands to the SDSS through the voice of the user, automatic navigation systems which direct a blind user of his possible moves, touch screen sensors, sensors which adjusts the display area automatically depending on the motion of the platform etc. In an SDSS, the GUI displays the decision options through various mechanisms such as graphical and tabular reports, 2-D and 3-D displays, bar charts, pie charts, scatter plots, line plots, application-specific plots and reports etc.

14.5 Work Flow of an SDSS

The work flow of an SDSS is given in Fig. 14.6. It shows the major steps involved during the decision making process of a problem involving spatio-temporal data.

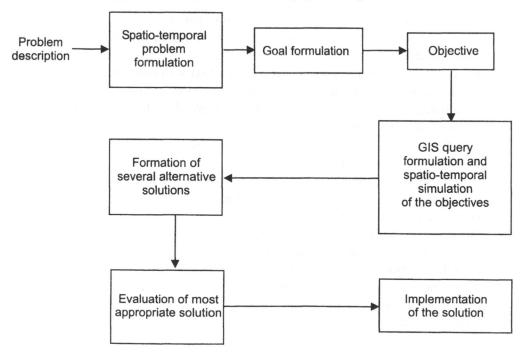

Fig. 14.6 Work flow of SDSS

14.6 Applications of SDSS

There are numerous examples of usage of GIS as a spatial decision support system. In the defense field, SDSS is primarily used in many C2 (command, control and intelligence), C3I (command, control, communication and intelligence) and C4I (command, control, communication, computers and intelligence) systems. SDSS assists the battle manager to make effective and alternate battle decisions; its role as operation planner, visualizer, analyzer etc. is a great asset to the battle commander (Densham 1991).

In an SDSS the spatio-temporal data of GIS are modeled using various domain rules. These rules can be used to classify terrain and identify the suitability of particular areas or locations for a specific operation. These areas are highlighted or visualized as possible alternatives for carrying out the operations on the terrain. Sometimes the decision to be taken depends on the spatio-temporal data in conjecture with tactical rules and non-spatial attributes of objects for which the suitability criteria is to be designed. Hence GIS is used as a collaborative platform for modeling, simulation and visualization involving spatial, temporal and non-spatial data (Enache 1994).

A scaled-down version of a DSS is a system, which provides DSA (decision support aids) whereby the system offers multiple decision options to a specific problem. It aids an expert to choose one out of many likely decisions to the problem.

Following are some important consideration to be taken into account while using GIS as SDSS.

a. Resolution and accuracy of spatio-temporal information.
b. The need for the tactical criteria/rule to be modeled in the form of a logical or mathematical formula.
c. The query criteria or the model under consideration with input patterns.
d. The mechanism of visualization viz. reports, charts or graphics e.g. 3D terrain visualization or depiction with the features highlighted on the background map etc.

Applications of SDSS includes

• Modeling and prediction of meteorological conditions
• Spatial distribution of demographic data
• Location allocation decisions regarding industrial set up, business set up etc.
• Decisions regarding allocation of water to irrigation and drinking purposes.
• Management of infrastructure development
• Monitoring and management of crime, epidemic in a locality etc.
• Decision regarding operations such as deployment of forces, weapons, sensors etc.

Some of the key benefits of using SDSS are:

1. Improving personnel and organizational efficiency
2. Expediting problem solving
3. Facilitating interpersonal communications
4. Promoting learning or training
5. Increasing organizational control

14.7 Examples of SDSS

Enlisted below are some important situations when SDSS involving terrain come in handy. These are merely a small sample of the application areas involving GIS—the actual numbers and exhaustive list is much more and ever increasing (Enache 1994).

- Identifying suitable areas for helipad construction.
- Identifying suitable location for deployment of guns.
- Classifying land for tank ability or identifying routes for cross-country mobility of tanks.
- Identifying the shortest, optimum and alternate routes among many possible routes between starting point and the destination point.
- Identifying suitable locations for the deployment of sensors, radars and communication equipments.
- Computing catchments areas and flood zones
- Identifying suitable locations for establishment of ammunition dump.

The implementation of these SDSS using GIS varies for different sets of spatio-temporal input data, user's criteria, set of tactical rules and forms of visualization (Keenan 1997). In the following sections, some of these examples are explained, elaborating on the candidate data set used as input, the processing steps used and the decision rules fired while obtaining the spatial decisions.

14.7.1 Identification of Sites Suitable for Construction of Helipad

INPUTS

Map/map board, corresponding DEM (digital elevation data), recent satellite image of the corresponding area.

PROCESSING/SEQUENCE OF RULES APPLIED

1. Choose the search area or area of interest (AOI) on the map/map board
2. Compute the slope polygons (C1), the aspect and the grid of height values of the AOI
3. Drape the satellite image on the DEM
4. Compute the slope and aspect of each point from the DEM
5. Filter all polygons having area > 300 × 300 sq meter areas (C2)
6. From the above polygons (C2) filter all polygons having slope < 15 degree (C3)
7. Filter all polygons (from C3) where the aspect < 3 from true north (C4)
8. Discard all polygons (from C4) having vegetation within this area (C5)
9. Sort the polygons (among C5) according to the height from MSL (least height from MSL)

If the soil characteristic of the polygon having least height has (hardness > 10) then among all such possible locations, the suitable location for the construction of a helipad is that which has the highest height from MSL (mean sea level)

OUTPUT

1. Prepare the 3D perspective view from the DEM
2. Highlight the suitable polygon on the 3D perspective view of the terrain with satellite image or raster map overlaid
3. Mark (with hatching) all possible patches on the map as well as the 3D view (Fig. 14.7)

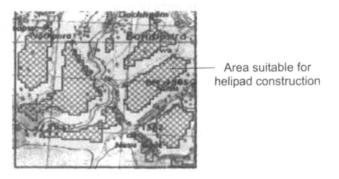

—— Area suitable for
helipad construction

Fig. 14.7 Suitable sites for helipad construction

14.7.2 Identification of suitable location for placement of guns

INPUT

Map/map board, corresponding DEM (digital elevation data), recent satellite image of the corresponding area.

PROCESSING/SEQUENCE OF RULES APPLIED

1. Prepare the 3D elevation model, drape the satellite image on it
2. Compute slope and aspect from the DEM, spot height and trig. height
3. Find the areas which has slope < 0.3 and aspect ≥ 0.000 and having no crest surrounding 400 m × 400 m and hardness of ground > 10
4. Choose areas closest to road, track, mule track

OUTPUT

Identifies and highlights suitable locations for placing air defense guns in a 3D perspective view of the terrain as depicted in Fig. 14.8.

Fig. 14.8 Suitable locations for siting AD guns

14.7.3 Flooding Simulation

Flooding simulation in a GIS helps the decision maker in comprehension and arriving at an intuitive decision. Some of the important decisions that can be arrived at and quantified are:

   a. computation of the volume of water that is required to flood an area of interest.
   b. computation of the time required to flood a particular area.
   c. identifying the suitable point of breach for optimum and fastest flooding.
   d. computation of the time and rate of flow required to maintain a certain level of flooding etc.

Some of these queries and decisions are important military considerations and some have application in civil areas (Lyon et al. 2000).

INPUT

Map/map board, corresponding DEM (digital elevation data), recent satellite image of the corresponding area, hydrographic survey data of the area and the soil condition

PROCESSING/SEQUENCE OF RULES APPLIED

   1. Identify the area where the effect of flooding is desired.
   2. Identify all the water resources such as rivers, tanks, canals etc. with their water capacity, flow direction and speed.
   3. Compute slope and aspect from the DEM, spot height and trig. height.
   4. Identify the ideal breach point and its dimension i.e. width and height.
   5. Compute the rate of flow of water taking into consideration the speed and direction of the flow of water.

6. Compute the flow of water into the flood zone taking into consideration the slope, aspect of each and every point of the flood zone.

7. Mark the progressive effect of flooding on the map.

OUTPUT

The effect of flood in time varies depending upon the amount of water discharged at the point of breach, the speed of water, the soil condition of the area under flooding etc. Given in Fig. 14.9 is a snap shot of a flood zone.

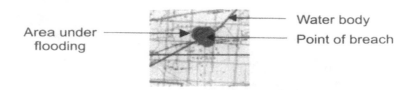

Fig. 14.9 Suitable locations for breaching the water body

References

[1.] Abel, D. J., P. J. Kilby and J. R. Davis. The system integration problem. *International Journal of Geographical Information Systems* 8:1–2. 1994.

[2.] Alter, S. L. *Decision Support Systems: Current Practice and Continuing Challenges.* Reading, Mass.: Addison-Wesley. 1980.

[3.] Andes, N. and J. E. Davis. Linking public health data using geographical Information System techniques: Alaskan community characteristics and infant mortality. *Statistics in Medicine* 14:481–90. 1995.

[4.] Anselin, L. and A. Getis. Spatial statistical analysis and geographic information systems. In *Geographic Information Systems, Spatial Modeling, and Policy Evaluation* edited by M. M. Fischer and P.Nijkamp. Berlin: Springer:35–49. 1993.

[5.] Batty, M. Using GIS for visual simulation modeling. *GIS World* 7(10):46–8. 1994.

[6.] Birkin, M., M. Clarke, G. Clarke and A. G. Wilson. *Intelligent GIS: Location decisions and Strategic Planning.* Cambridge (UK): GeoInformation International. 1996

[7.] Brimicombe, Allan. GIS, Environmental Modeling and Engineering. George Green Library. 2003.

[8.] Densham, P. J. Spatial decision support systems. In *Geographical Information Systems: Principles and Applications* edited by D. J. Maguire, M. F. Goodchield and D. W. Rhind. Harlow, Longman/New York: John Wiley & Sons Inc. vol. 1:403–12. 1991.

[9.] Druzdzel, M. J. and R. R. Flynn. Decision support systems. In *Encyclopedia of Library and Information Science* edited by A. Kent. Marcel Dekker, Inc. 1999

[10.] Easa, Said and Yupo Chan (eds.). *Urban Planning and Development Applications of GIS*. Reston, Va: American Society of Civil Engineers. George Green Library TD160 URB, 2000.

[11.] Enache, Miracea. Integrating GIS with DSS: A research agenda. URISA, pp.154–166, www internet page at http://wwwsgi.ursus.maine.edu/gisweb/spatdb/urisa/ ur94015.html, 1994.

[12.] Federa, K. and M. Kubat. Hybrid geographical information systems. *EARSel Advances in Remote Sensing* 1:89–100. 1992.

[13.] Finlay, P. N. *Introducing Decision Support Systems*. Oxford, UK Cambridge, Mass.,: NCC Blackwell; Blackwell Publishers. 1994.

[14.] Gachet, A. *Building Model-Driven Decision Support Systems with Dicodess*. Zurich: VDF. 2004.

[15.] Goodchild, M. F. A spatial analytical perspective on geographical information systems. *International Journal of Geographical Information Systems* 1:321–34. 1987.

[16.] Goodchild, M. F., R. P. Haining and S. Wise. Integrating GIS and spatial data analysis: *problems and possibilities. International Journal of Geographical Information Systems* 6:407–23. 1992.

[17.] Holsapple, C.W. and A. B. Whinston. *Decision Support Systems: A Knowledge-Based Approach.* St. Paul: West Publishing. 1996.

[18.] Keen, P. G. W. Decision support systems: A research perspective. In *Decision support Systems: Issues and Challenges* edited by G. Fick and R. H. Sprague. Oxford, New York: Pergamon Press. 1980.

[19.] Keen, P. G. W. *Decision Support Systems: An Organizational Perspective*. Reading, Mass., Addison-Wesley. 1978.

[20.] Keenan, Peter. Using a GIS as a DSS generator. Working Paper Management Information Systems 95-9, Michael Smurfit Graduate School of Business, University College Dublin, www internet page at http://mis.ucd.ie/staff/pkeenan/gis_as_a_dss.html, 1997.

[21.] Little, J.D.C. Models and managers: The concept of a decision calculus. *Management Science.*16(8): 466–485. 1970.

[22.] Lyon, John G. GIS for Water Resources and Watershed Management. Hallwood Library. 2000.

[23.] Maguire, D. J. Implementing spatial analysis and GIS applications for business and service planning. In *GIS for Business and Service Planning* edited by P. A. Longley and G. Clarke. Cambridge (UK): GeoInformation International. pp. 171–91. 1995.

[24.] Moore, J.H. and M.G. Chang. Design of decision support systems. *Data Base* 12: nos.1 and 2. 1980.

[25.] Power, D. J. What is a DSS? *The On-Line Executive Journal for Data-Intensive Decision Support* 1(3). 1997.

[26.] Power, D.J. A brief history of decision support systems. DSS Resources.COM, world wide web, version 2.8, May 31, 2003.

[27.] Skidmore, A. Environmental Modeling with GIS and Remote Sensing. Taylor & Francis. Hallward Library. 2001.

[28.] Sprague, R. H. and E. D. Carlson. *Building Effective Decision Support Systems.* Englewood Cliffs, N.J.: Prentice-Hall. 1982.

[29.] Steyaert, L. T. and M. F. Goodchild. Integrating geographic information systems and environmental simulation models. In *Environmental Information Management and Analysis: Ecosystem to Global Scales* edited by W. K. Michener, J. W. Brunt, S G. Stafford. London: Tayler and Francis. pp. 333–57. 1994

[30.] Warren, I. R. and H. K. Bach. MIKE 21: A modeling system for estuaries, costal water and seas. *Environmental Software* 7:229–40. 1992.

Question

1. What is the characteristic of a DSS?
2. What are various types of DSS?
3. What is a spatial decision support system (SDSS)?
4. Why is GIS a suitable platform for SDSS?
5. What is the architecture and sub-systems of a SDSS?
6. Give the work flow of a SDSS?
7. What is the difference between a DSS and an SDSS?

15

GIS: A Developer's Perspective

Once the decision has been made to use a GIS, the project manager then has a new set of decisions to make: 'What is the appropriate GIS for this project?'; 'What spatial data types are required to achieve specific GIS functions?'; 'What should be the configuration of the GIS?'; 'What are the spatial data formats supported by this GIS?'; 'Should GIS be used as a system, sub-system, or tool in an overall system?' etc. A project manager needs to have clear technological answers to all such questions at the beginning of the project. But there is no simple and clear answer to these questions. Often these queries lead to technology decisions which conflict each other. Sometimes, there is more than one technology choice available for a particular technology need. The manager has to take an appropriate technological decision for successful system realization. That is why, so far, this book has illustrated and analyzed various instances of GIS being used as a system, sub-system or tool for processing, analysis and visualization of geo-spatial data.

There is a persistent debate among mainstream information system developers and GIS professionals regarding the role of GIS. The heart of the debate is the role of GIS as a utility, tool or system in the overall system. Its capability in processing, analysis, visualization and interpretation of geo-spatial information accurately is also being questioned. An appropriate answer to the debate can be evolved by professionally analyzing the system requirements and quantifying geo-spatial requirements. Requirement analysis will put forward multiple options for different scenarios of GIS application. Software engineering (Pressman 2005) has a clear answer and processes for requirement engineering which forms the basis of application development in general, and GIS development in particular. Geo-spatial requirement specification is the prime driver of design and development of any spatial information technology project. Hence the quantum of spatio-temporal requirements should decide the role of GIS in the overall system (Antenucci et al. 1991).

This chapter starts with the pertinent question: 'How to choose an appropriate GIS?' Answering this question leads to determining the important criteria for choosing GIS and prioritizing them according to the requirement of the system. Therefore, in order to develop a system with GIS, the system requirements have to be analyzed. The next section discusses the process of software requirement specification. Depending on the quantity of spatial requirements and the system configuration, the capability and role of GIS have been established. The development approach that needs to be adopted in different scenarios has been arrived at. The role of software engineering principles and tools in quantifying software requirements and functional requirements has been discussed. The strategy to be adopted for successful implementation of GIS is recommended in the form of a flow chart exhibiting various decision options. Finally, three different scenarios for choosing GIS are described.

15.1 How to Choose a GIS?

Choosing a suitable GIS from many competing GIS products require a comparative study of their technology features. GIS products can be characterized by many attributes. To name a few, some important attributes are its functional capability, data handling capability, deployment architecture, re-sizing, ease of integration, customization etc. To arrive at a correct choice one must be aware of the multiple technology options available for these attributes and their influence on the system design. Hence the first step in evaluating an appropriate GIS is to make a comparative study of its attributes. Then these attributes have to be prioritized according to the need of the system to be realized. The prioritization of attributes and removal of conflict can be done using many time-tested techniques for technology evaluation. Prominent among them are

    a. analytical hierarchy process (AHP)

    b. decision aid for technology evaluation (DATE).

These multi-criteria decision analysis tools remove conflict and help in making apt technology decisions collaboratively.

Developing or customizing a GIS *ab initio* is different from choosing a GIS product from existing options. Customization of a GIS product requires that we follow the complete system development lifecycle. Following are some of the important issues that need to be objectively looked into while developing a GIS.

1. Geo-spatial requirement analysis
2. Functional requirement specification
3. Spatial input domain
4. Spatial processing capability
5. Output range
6. System architecture
7. Integration requirements
8. Cost and licensing policy

Different criteria leading to a successful implementation and choice of GIS need to be objectively evaluated and prioritized. The first step to achieve this is to analyze and quantify the system requirements. From these requirements, the specific requirements pertaining to geo-spatial requirements need to be apportioned. Geo-spatial requirements lead to an analysis of the geo-spatial functions and the spatial input domain required by the envisaged system.

To achieve the system goals, the system architecture which will decide the deployment of GIS components, needs to be designed. A thumb rule can be applied to decide on the role of the GIS and subsequently the lifecycle to be adopted to implement the requirements. The input domain, output range, I/O devices, information architecture and integration requirements can be derived from the requirement analysis. Therefore, the quantum of requirement is the prime driver for choosing the GIS components and the strategy of development.

15.2 Geo-Spatial Requirement Analysis

Requirement analysis is the first and crucial stage of a system development lifecycle (SDLC). Well established software engineering practices and tools exist to capture and quantify the system requirements (Pressman 2005). GIS requirements are in a way different from that of mainstream information system requirements because of the extra spatial coordinates attached with them. This calls for the analyst to look into aspects such as dimension, geometry, topology, computation and visualization associated with the spatial data. Each of these aspects calls for a specific design to perform the spatial functions. Hence spatial requirement analysis is tightly coupled with dimension, geometry and topology, which needs specific functions for measurement, computation, and visualization. A typical pattern of analysis which helps in examining spatial information is known as the model–view–control (MVC) method. A model–view–controller is an architectural pattern used in software engineering.

In designing complex computer applications that present and visualize a large amount of data to the user, a developer often wishes to separate data (model) from the user interface (view) concerns, so that changes in the user interface will not affect data handling, and data can be re-organized without changing the user interface. The model–view–controller solves this problem by decoupling data access and business logic from data presentation and user interaction. It does this by introducing an intermediate component: the controller. MVC plays an important role in analysis and design of GIS.

Another aspect of quantifying GIS requirements is its perception at various levels. An atomic requirement is one which performs a systemic function from a developer's perspective, whereas a unit requirement from a user's perspective can be one which performs an end-to-end work flow. One work flow of the user can be envisaged as more than one requirement by the designer. Therefore, the requirements at different levels of system analysis vary. Software engineering has different levels of system analysis and formal methods to measure and quantify the requirements. The effort, cost and time required to design and develop the spatial requirements can be estimated from the requirement specifications.

Using OOD (object-oriented design), the requirements are abstracted to use case (UC), a technique used to capture the functional requirements of the system. Then interaction of various processes is analyzed using sequence diagrams. A sequence diagram captures the details of interaction of the processes through messages and controls. Figure 15.1 depicts the modeling sequence of the GIS requirement analysis.

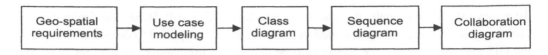

Fig. 15.1 Modeling sequence of a GIS requirement analysis

In the above process, the requirements have been modeled according to model–view–control (MVC) paradigms. It analyzes details of how the function will interact with the user, what will be its interface with the rest of the program, its interface with other systems and what are the class definitions which model the objects to be implemented. The functions, their internal and external parameters and how the functions will collaborate with the components of the overall system give a developer an idea as to how to estimate, design, develop and test the GIS.

15.3 Functional Requirement Specification

Functional requirement specification is an outcome of MVC modeling of spatial requirements. Functional specification of a GIS decides the capability or potential of the analytical competence of the system. Many a times, a geographical information system is benchmarked by the macro- and micro-functionality it offers. Generally functions are realized in the form of various geo-processing tools. These functions manifest in the overall system as menu options interacting with the user through the keyboard, dialog box or input/output (I/O) devices etc. Table 15.1 enlists some typical GIS functions with their API (application program interface) along with their likely inputs as argument to the API. The list is not exhaustive and can vary depending on a specific GIS application.

Table 15.1 Functional specification of GIS

Sl. No.	Macro- and micro-functions of a GIS
1.	**Vector/raster map processing and display functions** Read_and_display_raster_map_in(TIFF/ GIF/ JPEG/ PCX/ XWD) format Read_and_display_vector_map_in(DVD / DGN/ DXF/ Shape) format Read_and_display_DEM_in(DEM/ DTED) format Zoom/shrink/scroll/pan the displayed vector / raster map
2.	**Raster and image processing functions** Add_raster_map (map_name as string) Find_raster_format (map_name as string) Read_raster_map_header_info(map_name as string) Display_raster_map (map_name)

(continues...

Table 15.1 continued)

Sl. No.	Macro- and micro-functions of a GIS
	Adjust_image_brightness (value as integer [0–255]) Adjust_contrast (value as integer [0–255])
	Filter_noise (high pass/low pass) Adjust_linear_contrast (percentage as integer) Contrast_enhancement_image (image) using histogram equalization Detect_edges_in_image (name_of_filter) slobel / high pass / low pass Threshold_image(thres as integer) Thinning_of_image_edges (image) noise removed, edge detected image
	Negate_image (image_name) Rotate_image(image_name) Thematic_classification_of_image (image_name, list of themes, rules of classification) False_colour_composition_FCC (image) Principal_component_decomposition RGB (image)
	Zoom/shrink/scroll/pan the displayed image
3.	**Map control functions** Add_thematic_layers (spatial_layer_name) Highlight_thematic_layer (thematic_layer_name) Arrange_thematic_layers (vector_map, list_of_themes) Display_thematic_layer_at_top(thematic_layer_name) Display_thematic_layer_bottom (thematic_layer_name) Set_default_layers_for_display (list of thematic_layers) Set_display_on_or/off (thematic_layer_name, switch (on/off))
	Remove_thematic_layer (thematic_layer_name) Set_map_background__colour (R, G, B) Set_legends (thematic_layer_table) Set_line_style (thematic_layer_name, line_style) Set_width_of_display (display_width) Set_height_of_display (display_height)
4.	**GIS computation functions** Compute_location_of_object (object_id, lat, long) Compute_height_of_object (object_id, height) Compute_bearing_range_of_object (self_location, object_id) Compute_length (from_loc, to_loc) along road, cumulative, shortest in a road network, crow fly distance Compute_perimeter (set of points marked along the object) Compute_area (set_of_points surrounding the object) Compute_volume (set_of_points surrounding the object, depth) contours describing the piece of earth surface

(continues...

Table 15.1 continued)

Sl. No.	Macro- and micro-functions of a GIS
	Compute_azimuth_elevation (self_location, date and time, celestial object) Given_ENA_compute_LL (E, N, A) Given_LL_compute_ENA (latitude, longitude)
	Compute_slope_of_point_on_map (x, y, z) Compute_aspect_of_point_on_map (x, y, z)
	Compute_height_from_MSL (latitude, longitude) Compute_height_from_AGL (latitude, longitude)
	Compute_linear_distance ($x_1, y_1, z_1, x_2, y_2, z_2$) Compute_shortest_distance ($x_1, y_1, z_1, x_2, y_2, z_2$) Compute_optimum_distance ($x_1, y_1, z_1, x_2, y_2, z_2$) Compute_minimum_maximum_height_of_an_area (set_of_coordinates)
	Compute_perimeter_of_surface_area
	Compute_volume_of_earth_under_consideration Compute_differential_volume·of_surface Compute_thrust_vector (elevated_surface)
5.	**Computational Geometric Functions** Compute area_of_triangle ($x_1, y_1, x_2, y_2, x_3, y_3$) Boolean check_orientation_CCW($x_1, y_1, x_2, y_2, x_3, y_3$) Boolean intersect_lines ($x_1, y_1, x_2, y_2, x_3, y_3, x_4, y_4$) Boolean point_inside_triangle ($x_1, y_1, x_2, y_2, x_3, y_3$)
	Compute_convex_hull_of_planar_domain (set_of_coordinates) Compute_Voronoi_tessellation_of_planar_doman (set_of_coordinates) Compute_Delaunay_triangulation_of_planar_domain (set of_points) Boolean point_inside_polygon (point, polygon) Boolean point_inside_circle (point, circle) Boolean point_inside_sphere (point, sphere)
	Smooth_Delaunay_triangulated_domain View_tesselated_frame_of_terrain_surface View_trianglulated_irregular_network_of_terrain_surface Locate_observer and destn_on_TIN Compute_LOS_on_TIN Locate_point_on_TIN()
6.	**Spatial data manipulation functions** Sort_2D_coordinate_list (ascending_in_x / ascending_in_y) Search_coordinate (set_of_coordinates, coordinate_to_be_searched) Merge_map_files (file1, file2)

(continues...

Table 15.1 continued.)

Sl. No.	Macro- and micro-functions of a GIS
	Sort_3D_coordinate_list (ascending_in_x, ascending_in_y, ascending_in_z) Generate_ bounding_rectangle (set_of_points) Generate_ bounding_convex_polygon (set_of_points) Generate_bounding_sphere (set_of_points) Intersection of two lines in three-space and 2D space Intersection of three planes Render_3D_surface Simulate fog and haze to visualize terrain surface
7.	**Spatial data rendering functions** Add_point (x, y, z) Set_rectangle (x_1, y_1, x_2, y_2) Set_circle (x_1, y_1, radius) Triangular_surface_plot [xyz, viewpoint → {xv,yv,zv}] Triangulate_TIN (x y z) Find_contours (x y z, n levels, intervals) Generate_grid_from_TIN (set_of_coordinates) Find_display_device_parameters (x, y, w, h, display_buffer_depth, colour) Find_view_port_parameters (xv, yv, zv) Set_eye_coordinates (ex,ey,ez) Set_view_point (xv, yv, zv) Set_light_source_coordinates (xl, yl, zl) Drape_image_to_surface (image.tiff, surface) Generate_scene_from_surface (viewpoint) Generate_series_of_scenes_in_flt (surface) Fly_thru_the_ground Fly_over_terrain_surface Walk_thru_terrain_surface Compute_line_of_sight (observer's posn., object's posn.) Display_profile_view_of_the_terrain (P1, P2) Shaded_surface_with_almanac (viewpoint, almanac_data) Compute_ALMANAC_data (sun /moons posn (x, y, z) azimuth, elevation) Generate_contour_from_surveyed_data (xyz, interval, n levels) Generate_colour_coded_view (contours) Generate_shaded_relief_view (x y z, viewer's_posn. (R, è, ö), Sun_position (r, è, ö) $f[x\_, y\_]$ = Fit[xyz, polynomial, {x, y}] Plot_3D[$f[x, y]$, x_{max}, x_{min}, y_{max}, y_{min}, z_{max}, z_{min}, Scale_x, scale_y, scale_z, no_of_points, viewpoint [xv, yv, zv]

(continues...

Table 15.1 continued)

Sl. No.	Macro- and micro-functions of a GIS
8.	**Spatio-temporal query analysis** Connect_to_database (connection parameters to database) IP address of the DB server, name of the schema of the spatial database, login, password Prepare_buffer_query (feature_type, criteria) query string Prepare_attribute_query (attribute, feature_type) query string Prepare_spatial_query (coordinate_location) query string Execute_or_fire_query (DB_manager, spatial_query) result set stored in buffer Display_result_set_as_table (result_set) Display_and_highlight_result_graphically (result_set)
9.	**Map projection and transformation functions** Choose_map_data (map_name) Choose_datum (list_of_datum) Choose_map_projection_type (list_of_map_projections) Apply_map_projection (map_data, datum_type, map_projection) Display_projected_map (map_data) Apply_viewing_transformation (map_data, parameters_of_trans) Apply_geographic_transformation (map_data, reference_ellipsoid_parameters) Transform_device_coordinate_to_display_coordinate (è, ë, x, y) Transform_spherical_to_rectangular_coordinate (R, è, ö, easting, northing) Transform_rectangular_to_spherical_coordinate (R, è, ö, easting, northing) Display_map_2D(toposheet_no./ map_board_no/ place_name/ ENA) Display_map_3D(toposheet_no / map_board_no/ place_name/ ENA)
10.	**Spatial data conversion functions** Choose_source_map_data (source map format) Choose_target_map_data (target map format) Convert_map_data (source_map, target_map)
11.	**Symbol, overlay and map marking functions** Create_save_symbols_in_symbol_library (sym_id) Search_symbol (sym_lib) Attach_attribute_to_symbol (attribute_record) Place_symbol_on_map (lat/long/GR/..) Display_overlay (overlay_name) Save_overlay (overlay_name) Delete_overlay (overlay_name) Send_overlay (overlay_name, destination_user_id)

15.4 Spatial Input Domain of GIS

Input domain is an important benchmark criterion of GIS systems. Cardinality of the input domain is directly proportional to its functional capability and hence is a logical outcome of the requirement analysis and functional requirement specification. Different applications are developed around specific set of requirements which need different inputs to accomplish the requirements. Hence the quantity and type of spatial data to be processed by the system is derived from requirement and functional analysis of the GIS to be developed. The model–view–control (MVC) model helps in segregating data from the computation and graphical user interface (GUI). It helps in designing GIS where the impact of change in GUI does not have an impact on its input data and vice versa. The input domain of GIS is characterized by its data type, format, topology, dimension, co-ordinate system etc. A typical input domain of GIS describing the input data types and its associated functions is given in Table 15.2.

Table 15.2 Input domain specification of GIS

Sl. No.	Types of spatial data	Popular spatial data format	Typical function they perform
1.	Vector data	DGN, DVD, DWG, DXF, DXF, Arc, Shape, S-56	Thematic map composition e.g. communication map, administrative boundary etc. Position computation e.g. latitude, longitude, altitude; easting, northing, height etc. 2D measurements e.g. distance, area, perimeter etc.
2.	Raster data	TIFF, GIF, PCX, JPEG, PNG, BIL, BIP	Map display and printing, visualization and terrain change detection etc.
3.	Corner and boundary information	.bnd, .dwg, GCP-CP pair, .cnr	The upper left and lower bottom boundary coordinates of a rectangular topo-sheet (map) or an image. The groundc correlation points in case of satellite images. This kind of data is required to geo-dode and geo-reference an image or digital vector data.
4.	Geodatic datum and grid parameters	Standard parallel (SP), false easting (FE), false northing (FN), semi-major axis (a)	To impart a coordinate system to the image or vector data and then apply

(continues...

Table 15.2 continued)

Sl. No.	Types of spatial data	Popular spatial data format	Typical function they perform
		semi-minor axis (*b*), eccentricity flattening parameter (*f*) etc.	proper map projection for correct spatial positioning.
5.	Elevation data	DTED (dt_0, dt_1, dt_2), DEM (.dem)	To display the undulated terrain surface in the form of DEM, for 3D perspective view, shaded relief view, fly through simulation or to measure height, volume etc.
6.	Non-spatial data	TIGER, Oracle dump (dmp) or mdb	Non-spatial attributes are the statistical data or the detailed recorded information about digital objects.

15.5 Spatial Processing Capability of GIS

The processing needs of a GIS are the direct outcome of its functional specification. A group of spatial processing tools in a GIS is often known as the processing engine of the GIS. The processing engine constitutes intelligent implementations of various algorithms such as map transformation, shortest path computation, geodesic computation, computational geometric algorithms, numerical analysis, spatial statistics etc. The specific processing needs of the system which will transform the spatial input domain to intermediate results and finally to the output range of the GIS need to be implemented while developing a GIS or should exist in the GIS system to be adopted for customization. The broad category of processing that a GIS system has is shown in the hierarchical diagram given in Table 15.3.

Table 15.3 Different category of GIS analysis

GIS analysis			
2D analysis	**3D analysis**	**Visual analysis**	**Simulation and SDSS**
Point Computation of E, N, Φ, λ, h, H	**Height** Computation of height from MSL	Visual display of 3D terrain in Perspective view	Fly thru simulation of terrain Terrain walk thru

(continues...

Table 15.3 continued)

2D analysis	3D analysis	Visual analysis	Simulation and SDSS
Given an object, to find coordinates of its location Given a coordinate, to find the object	Orthomorphic height given (F, l) of a point on the earth surface	Orthogonal view Colour-coded elevation model Shaded relief view	Terrain see thru Terrain profiling Wire Frame model Identification of suitable location/ area for a particular purpose
Line/distance Computation of distance, distance along terrain surface, optimum distance, aerial distance, cumulative distance	Computation of line of sight (LOS) between two points of earth surface	3D walk thru in terrain surface Fly thru and see thru over a terrain surface Computation of slope, aspect, surface area	Study of land cover, land use, water shed analysis and impact of disaster etc.
Polygon/area Computation of surface area, perimeter, normal area Given an object, to find which area/polygon it lies in Given an area/polygon, to find what are the objects inside them Query Spatial, attribute, hybrid, iterative, buffer zone	**Volume** Computation volume of a piece of undulated land / water body etc.		Fly thru simulation of terrain Terrain walk thru Terrain see thru Terrain profiling Wire Frame model Identification of suitable location/ area for a particular purpose Study of land cover, land use, water shed analysis and impact of disaster etc.

15.6 Architecture Issues of GIS

Software architecture decides the availability of the system resources and services to the user. While the system architecture decides important aspects such as scalability, throughput, reliability and security of geo-spatial data and processing. Therefore architecture plays an important role in choice of GIS. With GIS emerging as a collaborative information system, availability of geo-spatial computation and information across a host of applications and users in a distributed network becomes necessary. GIS data, processing and software can be available for distribution through the system constituting hardware, software and networking. Therefore system architecture is a holistic consideration of hardware, software and information architecture. GIS system designers must understand every aspect of the architecture before choosing a COTS GIS.

To meet the holistic requirements of such systems, major GIS developers have developed a set of GIS software suitable for specific application areas and adaptable to different architecture. These sets of GIS applications are called GIS suits.

Geo-spatial information has a high strategic and economic value and hence is susceptible to unauthorized usage. This necessitates a comprehensive security policy involving hardware, network, and communication; a system to guarantee information security has to be designed and implemented in GIS systems because of this. Architecture (hardware, information and software architecture) guides the security policy to be designed and adapted for such a system. Essentially system architecture consists of (1.) hardware architecture, (2) software architecture and (c) information architecture

15.6.1 Hardware Architecture

Hardware architecture gives information about the deployment of different hardware components e.g. number of GIS servers, client terminal, special display devices such as primary display, secondary display or dual display etc—sometimes, the display is in a hand-held terminal or head on display (HOD). The typical hardware configuration of GIS system is given in Table 15.4.

Table 15.4 Hardware configuration of GIS system

Sl. No.	Hardware component	Typical configuration
1.	Processor.	Dual CPU
2.	Memory	More than one GB primary memory and RAID (redundant array of discs), High display memoryHard disc encrypted mechanism.
3.	Input devices	Mouse, 3D mouse, stylus, scanner
4.	Output devices	Dual monitor display, high resolution plasma display, back light enabled hand-held display,head mounted display etc.Mouse, stylus, cyber globe and flock of birds (parts of a system following fairly simple rules and interacting with each other to form a cohesive and dynamic whole) Plotter and high resolution printer
5.	Display device	Special display driver with on-board display memory for renderings of scenes during fly through and DEM renderings. Head on displayData wall controlled by a central server

15.6.2 Software Architecture

GIS software architecture defines the software environment in which the system will be deployed and positioned so as to process the request and respond to users with optimal response time. A J2EE component architecture for an enterprise GIS system is depicted in Fig.15.2 with different software components deployed in an 3-tier architecture.

15.6.3 Information Architecture

Information architecture (IA) is the art and science of expressing a model or concept for information processed, generated and disseminated by the system. Information architecture is about where and how the flow of information is guided among the software components. It also defines the flow of request and response executed in the information cycle while the user interacts with the system. Hence information architecture guides the deployment and interaction of various software components in the overall system. A typical component and information flow architecture of a distributed enterprise GIS system is depicted in Figs 15.2 and 15.3 respectively.

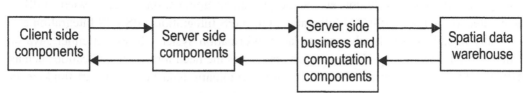

Fig. 15.2 Components and their interaction in a 3-tier architecture

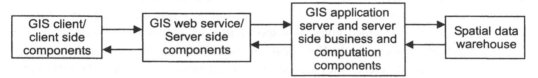

Fig. 15.3 Request response cycle in a 3-tier architecture

As mentioned in Chapter 1, depending upon the architecture, GIS systems are classified as desktop, client–server and enterprise information systems. These systems are different in terms of their servicing capability, deployment architecture, hardware and software configuration. Information architecture of GIS plays a crucial role in deciding its deployment in the networked computing environment. It should be decided after determing the number of clients using the system, their physical and logical distribution etc.

15.7 Integration Issues of GIS

GIS system integration is the process of bringing together the components of a GIS with other sub-systems as one overall system and ensuring that the sub-systems function together as a system. In information technology, systems integration is the process of linking together different computing systems and software applications and hardware to achieve the functionality of the overall system. In a sense, the system integrator brings together discrete systems utilizing a variety of techniques such as computer networking, enterprise application integration, in-house developed software components, work flow integration and business process management to achieve a cohesive integrated system.

Therefore GIS system integration involves integrating existing GIS components to other sub-systems. This can be achieved through development of sub-system interfaces capable of exchanging data and control with the interfaces of other sub-system interfaces. Integration involves joining the sub-systems together by 'gluing' their interfaces together.

15.7.1 Need for Integration

The application domain of GIS is increasing because of the wider applications of geo-spatial information in various walks of life. More and more systems and appliances use GIS software as GUI or interface to interact with the user, to analyze and visualize organizational information for spatio-temporal distribution etc. Some prominent application systems, where GIS is integrated as GUI are GIS-CAD Integration, GIS-GPS Integration, GIS-OIS Integration etc.

Many organizations already operate large information systems based on either Java or Microsoft.NET technology. If the company perceives the need for a new application such as GIS to add to their current technology then the automatic tendency is to start thinking in terms of current technology and integrating the GIS application with the existing application. Traditionally information systems are designed for non-spatial data and hence use relational data models. The user query in these information systems are formed through well-established structure query language (SQL) or QUALs. To extract information, the SQL query is fired to the information system, which gives its response to user queries in the form of textual reports, statistical graphs or charts. But in case of spatial information systems, the nature of data and its usage are different—they are mostly visual in nature, in the form of digital maps, charts or models. Also the kinds of query spatial systems can handle are far more improved, for example, GIS can handle spatial queries, temporal queries, buffer queries, attribute queries etc. A substantial part of the query is formed by a logical combination of the geometric predicates. These queries differ in their construction and resultant outputs. Spatial queries are mostly formed with visual tools and the results are reflected visually with a map as the background. The manifestations of these different queries are mostly in the form of an SQL having geometrical computation as part of the query. One can categorize all these information systems as an operation information system (OIS). Hence integrating GIS with already existing information systems is gaining popularity.

The integrated product is important because of its improved provision for comprehending information, i.e. the traditional textual reports of OIS can be depicted as a graphical report displaying the spatio-temporal distribution of the phenomenon under consideration. The spatio-temporal distribution and graphical representation of traditional reports have many advantages in comparison to the textual report. The analyst can make use of operation research techniques and common sense to derive better results from such visual display.

GIS and OIS are traditionally developed differently. They also have different input domain, processing capability and output range. Hence integrating them directly as a monolithic product or system where both the applications are compiled together and the compiled output manifest itself, as a single executable system is not possible.

The integration is a non-trivial effort and can be achieved in three logical steps.

STEP 1: Inter-operability among GIS and OIS i.e. obtaining spatial data inter-operability, software component inter-operability and control inter-operability

STEP 2: Interface i.e. designing the interface using XML/GML web services, runtime bridges, middleware messaging services, shared database, integration broker

STEP 3: Integration i.e. deploying the integration protocol appropriately in the architecture among heterogeneous software and performing integration test so as to achieve the system objective

The subsequent sections of this chapter further elaborates the above three steps and delve upon their importance while designing the GIS system.

All problems of integration of GIS to any other software can be mapped to three generic scenarios:

1. Integrating an already developed application to a newly developed GIS
2. Integrating a new application to an existing GIS
3. Integrating an existing application to an already developed GIS

The first scenario is the easiest but least common in practice, whereas the last is the most difficult and the most common in system design. There are a number of best practices recommended by industry for different scenarios but there is no generic approach to all the three. Hence practitioners and system integrators must judiciously adapt and evolve these integration scenarios during the early phase of the analysis.

15.7.2 Inter-Operability

The most frequent scenario of GIS-MIS integration is inter-operability. Inter-operability is the ability to communicate or transfer data between functional units running on different platforms, implemented in different technologies, using industry standards or widely accepted data description and communication protocols.

For example, if systems developed in J2EE and .NET platforms need to be integrated, then to ensure that the application built on one platform connects to those created on the other, inter-operability of data and control need to be developed. The industry has evolved different techniques to ensure inter-operability at different tiers of application. There are a number of ways in which inter-operability and integrations can be implemented. Enlisted below are some of popular integration technologies practiced in the industry:

1. XML/GML web services
2. Runtime bridges
3. Middleware messaging services
4. Shared database
5. Integration broker

Table 15.5 summarizes various inter-operability scenarios that can be implemented using the above integration technologies in different labels of applications.

Table 15.5 Different inter-operability scenarios

Integration/ Inter-operability technology	Presentation to presentation	Presentation to business	Business to business	Business to data
XML/GML protocol in web tier		Yes	Yes	Yes
Runtime bridges	Yes	Yes	Yes	
Messaging		Yes	Yes	
Shared database	Yes			Yes
Integration brokers		Yes	Yes	

Depending upon the operating scenario and system configuration, appropriate inter-operability strategy can be adopted to integrate applications. In the context of integrating a GIS with an application (say OIS, operational information system) developed in other technology than that of GIS, the inter-operability is achieved by performing the following logical steps.

1. Spatial data inter-operability (translation of spatial data used by OIS to native format of the GIS)
2. Encoding of non-spatial data emerging from OIS (using structured string/XML/GML)
3. Parsing and decoding of non-spatial data in GIS
4. Graphical representation of decoded non-spatial data in GIS (Graphic2Text)

The above logical steps have been depicted by a block diagram (Fig. 15.4) which depicts the interoperability constructs between GIS and OIS.

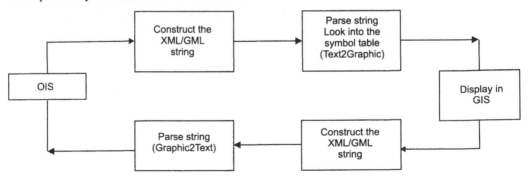

Fig. 15.4 Various stages of OIS–GIS integration

15.7.3 Recommendations for GIS–OIS Integration

There are a number of recommendations and mechanisms that aim at solving such problems. The XML serialization for Java and .NET platform which converts data type into an XML stream, converting it to an XML document, parsing functions to fragment data in the other end by a de-serializing function is a very good solution for such problems. But it suffers from the following restrictions and deficiencies:

1. The XML document under de-serialization contains fields, the serialized cannot process.
2. The object is not declared as a public object.
3. The object must contain a valid no-argument constructor.

The above deficiencies may lead to a failure in integration. Hence a generic protocol is to be developed where the data created by OIS using J2EE use cases, components such as JSP, BEANS etc. and stored in back-end database ORACLE need to be visualized in a geographical information system which is implemented using .NET platform (ASP/COM/DCOM and activeX components) (Mackenzie 2002). Visualization consists of both graphic visualization in terms of an appropriate symbol located at the spatial coordinates on the map and the attached attributes which are textual and stored in the back-end database. In a way it is the graphical depiction of the records representing ground features deployed dynamically on the map. Generally the static features of the earth are well surveyed and are part of the digital map, which is displayed in the background. The dynamic situation of any event e.g. the mobile fleet, or order of the battle (ORBAT) with the position of all the field units etc. which are continuously changing in their spatial location or non-spatial data need to be depicted as an overlay with the map as background.

While designing such protocol, in addition to the above scenarios and deficiencies, the following constraints need to be taken into consideration.

1. The data is a mix of spatial and non-spatial attributes.
2. The data need to be validated before sending and or receiving at the either end.
3. The data can be edited or not depending on its access rights.
4. The protocol must be independent of the data and control of different software platforms (e.g. .NET, J2EE etc.)
5. The meta data needed to exercise control over the data should be inbuilt.

The protocol explains how to construct the sending and receiving of a string; how to parse the string; tokenize it; and finally initialize the data to variable or data fields in respective systems.

The protocol consists of three parts:

1. The effective definition or format of the data to be exchanged as string
2. The mechanism to parse the string for tokenizing
3. The mechanism to exercise control over various fields

15.8 Cost and Licensing of GIS

Licensing mechanism is an important aspect of any application software as it directly defines the cost, scope and availability of services of the application. This is more important for GIS which allows controlled usage to the authorized user because the spatial data and information has high commercial value as well as strategic importance. Different licensing mechanisms have been evolved over the years, depending upon the usage pattern, pricing, distribution restriction, and purpose of usage by the industry.

Generally, application software comes with a license which specifically states the terms, and conditions under which the software may be legally used. Licenses vary from product to product and may authorize one computer or user to use the software or several thousand network users to share the application through the network to the user. It is important to read and understand the license accompanying the program to ensure that you have sufficient legal copies and are adhering to all terms of the license. Remember that when you purchase a software package you are in fact only acquiring a license to use it. The publisher retains full rights and has sole rights of distribution and reproduction.

Special licensing conditions are also applied according to the usage of the application software. For example, GIS used for educational or academic research have special conditions attached to them. These special conditions such as (a) administrative use only, (b) outside an administrative boundary specified by a condition, (c) usage for commercial purposes only, (d) usage for defense purpose etc. Different licensing policies adopted by COTS software are given in Table 15.6.

Table 15.6 Different licensing policy adopted by GIS

Sl No	Type of license	Explanation
1.	**Shrink-wrap license**	In the case of mass marketing of software products, the license agreement often takes the form of a shrink-wrap license. It is so called because a package containing the product and printed license agreement is shrink-wrapped. The legal basis is that, instead of agreeing to its terms by signing a written document, the customer signifies their agreement by conduct i.e. by opening the diskettes or using the software. These types of license are less practiced in case of GIS software.
2.	**License per machine**	Software licensed per machine requires that customers purchase a license for each PC that might use the software. Generally the license is tightly coupled to the signature of the computer, which is generally the CPU ID, hard disc serial number or the network ID. The GIS are mostly licensed per machine basis.
3.	**License per individual/ personal license**	Software can be licensed to a particular individual meaning that only a specific person can use it. This type of license is most suitable for software that will be used by only one person and at other times is idle in the system. They are in general very difficult to administer. This licensing policy is rarely used in GIS.
4.	**Concurrent license**	A concurrent license allows a limited number of users to connect simultaneously to a GIS application. For example, if you have 25 users on a network but only 10 use a GIS at any given time, then you would only need to purchase 10 licenses. The enterprise version of GIS uses this licensing policy.

(continues...

Table 15.6 continued)

Sl No	Type of license	Explanation
5.	Network license	Sometimes the license of a GIS is valid in a network of computing machines interconnected through a LAN/WAN. It is generally limited to a LAN/WAN or individual file server serving the LAN. How you or I define a network and how the suppliers define a network may be very different. It is important to read the terms of the license very carefully to determine the publisher's intentions. Every member of the network is normally allowed access. Many popular GIS servers provide service through this licensing policy.
6.	Site license	Site license is usually unlimited but with a time limitation. The term 'site' must always be defined. It can vary from a small department to a University network and its allied institutes. They are deployed in academic or scientific research environment.
7.	Shareware	This is software which is marketed by freely distributing a limited or full version of the software through trade shows, bulletin boards, and websites or by handing down from one user to another. Potential purchasers are encouraged to copy the program as a preview. The rule is that if you like the product and keep it, then you send the developer payment for it. If you keep it but do not pay for it you are in violation of the copyright. Almost all Shareware includes a read me file or an opening menu stating that the program is Shareware and how and where to send payment. BEWARE non-compliance.
8.	Freeware or public domain license	Distributed in the same way as Shareware but with no requirement for payment.
9.	Suite license	A suite is a group of applications sold together. Though a suite contains different applications, it is intended for only one user and the suite contains only one license. Applications within the suite may not be used concurrently i.e. for an Office license you may not install Excel on one machine and PowerPoint and Word on another. All applications in the suite must be installed on the same machine. Generally every COTS GIS has a suit or family of GIS products e.g. ESRI's Arc, Intergraph's Geomedia or Bentley suit of GIS has different sub-products bundled together.
10.	Upgrades/ updates	When upgrading, the upgrade is an improvement of the original application software—it does not create a second license. If you wish to retain the use of the previous version also, then two licenses are required.

(continues...

Table 15.6 continued)

Sl No	Type of license	Explanation
11.	**Competitive upgrades**	Many suppliers have created competitive upgrades. For example, you currently use one version of the GIS and wish to upgrade to the next version necessitate a separate license. In this case you still have two separate licenses and may continue to use both versions of the products.

15.9 Development Lifecycle of GIS

Systems Development Life Cycle (SDLC) or sometimes just (SLC) is defined as a system development process and is independent of software or other information technology considerations. It is used by a systems analyst to develop a GIS which includes requirements analysis, validation, testing, training, and user ownership through investigation, analysis, design, implementation, and maintenance. SDLC is also known as information systems development or application development. An SDLC helps in the production of a high quality GIS system that meets user's expectations, within time and cost estimates. SDLC for GIS is a systematic approach to problem solving and is composed of several phases, each comprised of multiple steps. The SDLC for GIS depends upon three main scenarios:

a. When GIS is used as a tool

b. When GIS is used as a sub-system of an overall system

c. When GIS is customized for a particular application.

SDLC for GIS to be used as a tool in an overall system includes:

(a) Implementation; (b) testing; (c) evaluation

SDLC for development of GIS from *ab initio* involves

(a) Feasibility study; (b) analysis; (c) design; (d) development; (e) testing; (f) implementation; (g) maintenance

SDLC for choosing a COTS GIS and further customizing and integrating it with a main system involves:

(a) Feasibility study; (b) technology evaluation; (c) analysis; (d) design; (e) customization; (f) testing; (g) evaluation; (h) maintenance

The SDLC for these three scenarios are further elaborated in the subsequent sections.

15.9.1 Scenario I (GIS as a tool)

If the geo-spatial requirements are less than 10% of the system requirements then GIS can be used as a tool, utility or component of the main system. Such requirements are mostly visualization of terrain information such as a map, DEM or a model of the earth surface. In most cases, the information pertaining to the mainstream system has to be depicted graphically against a map in the background. A popular example of such requirements is the display systems of a modern aircraft in the cockpit or the main GUI of the weapon system depicting the self-location and the target location in the backdrop of a map. In most such cases, the GIS is embedded in the main system serving as part of the main system. If the geo-spatial requirement is meant for computing or calculating the geographic measurements such as position, length, area, perimeter of the object on the earth then the GIS can be used as a separate utility fulfilling the requirement and the computed output can be fed to the main system as an input. If the GIS is used in this way, as a component or utility, then the process can be kept light.

15.9.2 Scenario II (GIS as a Sub-System)

If the geo-spatial requirements are more then 10% and less then 50% of the system requirements, then GIS can be developed as a sub-system of the main system. A recommended path of developing a GIS sub-system where the project lifecycle is small and the organization does not have a competence in GIS, is to identify a suitable COTS GIS, customize it to satisfy the system requirements and integrate the customized system with the main system.

Customization of GIS

Customization of GIS involves modification of the COTS GIS software using the library of the APIs provided in the COTS. Customization (Maguire 2000) essentially involves the base software configured to comply the input domain the application wishes to process and adapt various work flows the user wishes to realize. Mostly, the customized software needs to be integrated with in-house developed software. Generally there is no generic APIs available in the COTS as well as the in-house developed software to exchange both data and control back and forth from external software. In the absence of such data and control exchanging mechanism, integrating heterogeneous software modules, becomes non-trivial. This inter-operability problem of incompatibility of in-house developed software with COTS software becomes more acute if the software is in different platforms and in an enterprise environment.

COTS GIS applications are designed, developed and deployed for generic requirements. A few of the generic requirements are
1. Standard input geo-spatial data domain
2. Deployment architecture (compatibility with popular software and hardware platforms like OS, programming language, RDBMS etc)
3. 2D/3D analysis and visualization
4. Specific interfaces to popular products for presentation etc

There is a growing trend to reduce development cost and time for military system implementation and its deployment. This has resulted in an increasing demand for software customization of existing commercially off the shelf (COTS) products. Since GIS is an important component of any military Tac C3I system, customization of a COTS GIS is attempted. This involves identification and adaptation of spatial and non-spatial input data domain, design of user work flows, design of a protocol to inter-operate between the customized GIS and other sub-systems etc.

Important consideration in choosing appropriate COTS GIS to customize the requirements of the overall system are as follows.

1. Compatibility of the geo-spatial input domain i.e. the software should read, generate and display spatio-temporal data such as vector, raster, DTED (digital terrain elevation data) and other forms of spatial data in different format and scale, without loss of information.
2. Portability of the customized component from the COTS GIS application with the software and hardware environment of the mainstream application.
3. Compliance to the hardware, software, communication and information architecture of the main system.
4. Inter-operability with the main system in representing the information without loss of visual and textual context. This amounts to correct representation of cartographic and topographic data and user data in graphical form.

Often COTS GIS are distributed with a licensing policy for commercial utilization. The various licensing policy are given in Table 15.6.

The life cycle used to achieve customization of a suitable COTS GIS to meet the system requirements are:

1. Quantification of functional requirements
2. Evaluation of suitable COTS GIS using multi-criteria decision analysis or analytical hierarchical process
3. Select the COTS on the basis of T1L1 (technologically number one but lowest in cost).
4. Model the requirements using MVC pattern and DIKD (data–information–knowledge–decision) techniques and customize the work flows.
5. Develop suitable test stubs and test drivers to test these work flows. Test individual functions and work flows and the COTS component using the stubs and drivers.
6. Integrate the customized system to the parent system.
7. Test the integrated system for performance, field-testing and for user acceptance.

15.9.3 Scenario III (GIS as a System)

If the geo-spatial requirements are more then 50% of the system requirements then GIS can be developed as a system in itself and the other requirements of the system as supplements to the GIS. The system can be envisaged as a GIS-centric system. A *ab initio* approach in developing a GIS system is recommended. Before embarking upon such a venture, project feasibility study and organization capability are studied completely. A decision process such as decision aid for technology evaluation (DATE) analysis is important in this respect. A DATE analysis brings out the strength and weakness of the organization in technology, sub- and sub-sub technology involved in developing the GIS. Estimation is done to find the feasibility index for taking such a project and this helps in scheduling and tracking of activities and resources of the project.

The lifecycle adopted to start *ab initio* development of GIS is given in the following sequence of steps:

1. Requirement engineering (capture, analysis and quantification, prioritization of the requirements)

2. Decision aid for technology evaluation (DATE) analysis of the contributing and sub-technologies pertaining to the project including the integration strength
3. Preparation of Project feasibility report
4. Project estimation and scheduling (setting goals and milestones, review schedules, cost estimation and time estimation)
5. Design, development, test, integration and fielding

The three design scenarios are put in the form of a flow chart depicted in Fig.15.5. The recommended development paths for the three scenarios are depicted in the form of block diagrams (Fig.15.6). A developer or technology manager needs to adapt one of the development paths to meet the system requirements. In actual practice, a variant suitably adapted by a team is put into practice.

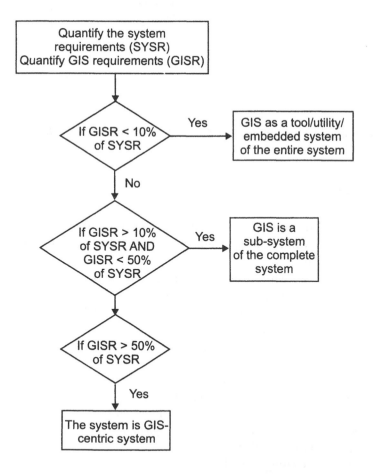

Fig. 15.5 GIS design according to quantity of GIS requirements

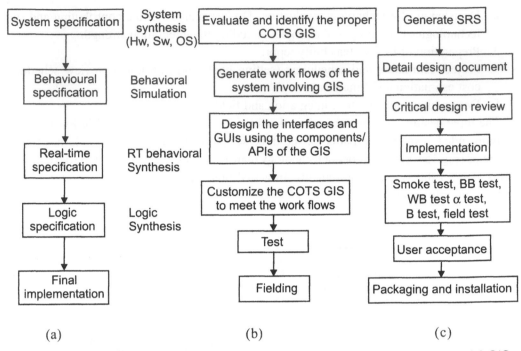

Fig. 15. 6 Design of GIS as an (a) embedded tool, (b) customized sub-system, (c) GIS-centric system

References

[1.] Maguire, D. J. GIS customization. In *Geographical Information System, Principles & Technical Issues*, vol. 1, 2nd ed., edited by Paul A. Longley et al. 359–369, John Wiley & Sons Inc. 2000.

[2.] Antenucci, J., K. Brown, Croswell and M. Kevany. *Geographic Information System – A Guide to the Technology*. New York: Van Nostrand Reinhold. 1991

[3.] Bell, D., I. Morrey and J. Pugh. *Software Engineering: A Programming Approach*, 2nd ed. New York: Prentice-Hall. 1992.

[4.] Escobar, F. Vector Overlay Processes, Sample Theory. The University of Melbourne. 1998.

[5.] Burrough, P.A. *Principles of Geographic Information System for Land Resource Assessment*. NY: Oxford Science Pub. 1986.

[6.] Sharkey, Mackenzie. *Teach Yourself Visual Basic .NET in 21 Days*. Techmedia. 2002.

[7.] Berry, J.K. *Beyond Mapping: Concepts, Algorithms and Issues in GIS*. Fort Collins, USA: GIS World Books. 1995.

[8.] Keith, Clarke C. Getting started with geographic information system, edited by Paul A. Longley et al. John Wiley & Sons Inc. 2000.

[9.] Roger S. Pressman. *Software Engineering: A Practitioner's Approach*. McGraw-Hill International Edition. 2005.

Questions

1. How to choose a GIS for a target application?
2. What are the important criteria a GIS analyst must have while designing a system involving GIS?
3. What are sequence flows for capturing GIS requirements?
4. Give the categories and sub-categories of GIS functions?
5. What are the components in a 3-tier information architecture?
6. What are different types of licensing policies in a commercial GIS?
7. What are different scenarios of GIS development life cycle?
8. Explain various stages of development of a GIS as a
   a. Tool
   b. Sub-system
   c. System

Appendix A

Analysis of GIS Aspects

To understand a geographical information system (GIS) it is pertinent to analyze various perspectives (planner's, owner's, designer's, builder's and sub-contractor's) of the system and answer the what, how, where, who, when and why of each of these aspects. A templet has been adopted to present the various aspects of the system with respect to the major influencing factors. To ensure that the study is complete and correct, a conglomeration of two software analytical methodology namely RM-ODP and Zachman Framework propounded by John A. Zachman has been adopted. The various perspective and aspects of GIS have been mapped to a matrix depicted in Table A.1. The rows of the matrix define various perspectives of GIS and the columns, various aspects influencing GIS. Hence each cell of the matrix reflects various aspects of GIS with respect to one of the influencing factor or aspect of the system. The mapping may not be exact and lucid because the small space of each cell makes it impossible to accommodate detailed information.

To analyze an enterprise information system like GIS, it is pertinent to understand the five views namely, enterprise view, information view, computational view, engineering view and technology view of the GIS. These views are influenced by six aspects of the system namely data, function, network, people, time and motivation. Studying the various aspect of GIS with respect to a particular view, answers many 'why' and 'how' questions of a GIS developer.

Perspectives (row)/aspects (columns)	Data What	Function How	Network Where	People Who	Time When	Motivation Why
Scope	Processing and computation of spatial, temporal and attribute data pertaining to objects of earth	Geo-processing or spatial processing tools applied over the spatial data	Stand-alone spatial information system	Survey agencies, Cartographers	Survey map making	It is a geo-processing tool box to answer the spatio-temporal queries
		Cartographic modeling of spatio–temporal data	Client–server system serving spatial information	Photogrammetrists	Tac C4I and C3I systems for battle planning and management	To automize map making and map updation
				City planners		
Planner		Computational geometric operations and techniques applied on the spatial data	Layered information architecture processing spatial information (3-tier architecture)	Space researchers		To track real-time movement geo-spatially
				Defence personnel and battle managers		To identify the input domain, sources and agencies of data
Enterprise model (conceptual)	Data model and data flow Raster data Vector data DTED data GPS data Attribute data Sat img → Geo-coded raster img → R2V conversion → Vec data attributization → composite spatio-temporal data	Map publishing with simultaneously viewing of a map for collaborative planningMap chat/ map conversation	Stand-alone WS for map views	Work flows of GIS data modeling (cartographer) Surveying (surveyor) Pre-processing, display and view (GIS analyst)		
			Spatio-temporal query and search terminals			
Owner		Map services to various clients through map server	Hand-held device and palm tops for survey and location in a networked environment	2D and 3D computation, reports and views (GIS user)		

Perspectives (row)/aspects (columns)	Data What	Function How	Network Where	People Who	Time When	Motivation Why
System model (logical)	Input Domain of GIS Physical data arch Raster data format Vector data format DTED data format	Back-end map data base Middle-tier map server (concurrent map server) Thin clients accessing and processing on the client end (3-tier architecture) map	GIS for stand-alone system GIS deployed in client–server mode	Geo-spatial data base designer Spatial database build and optimization Meta data creation		
Designer						
Technology model (physical)	Data definition Raster data (Header + Data + World Coordinate + Meta Data) Vector data (Spatial data + Meta data + Geo-coding information)	Decode and display map files Compute location in latitude, longitude and altitude, altitude. Compute distance shortest path, bearing 3D perspective view Fly through Walk through models		Spatial database build and optimization Meta data creation Spatio-temporal data base Spatial technology Query builder Spatial query Attribute query Buffer query Iterative query Hybrid query		
Builder						
Detail Presentation (Out-of-Context) Sub Contractor						
Function Enterprise	Data	Function	Network	Organization	Schedule	Strategy

Table A.1 RM-ODP analysis of GIS aspects

Appendix B

Glossary of Definitions and Terms

1. **Accuracy**   The closeness of observations, computations or estimates to the true value as accepted as being true. Accuracy relates to the exactness of the result, and is distinguished from precision which relates to the exactness of the operation by which the result was obtained

2. **Address**
   1. A means of referencing an object for the purposes of unique identification and location
   2. The location of a block of computer memory

3. **Attribute**
   1. Descriptive information characterizing a geographical feature (point, line, area)
   2. Commonly, a fact describing an entity in a relational model, equivalent to the column in a relational table

4. **Bleeding edge**   That edge of a map or chart on which cartographic detail is extended beyond the neat line to the edge of the sheet.

5. **Cadastre**   A public register or survey that defines or re-establishes boundaries of public and/or private land for purposes of ownership and/or taxation

6. **Cartography**
   1. The organization and communication of geographically related information in either graphic or digital form; it can include all stages from data acquisition to presentation and use
   2. The production of maps, including construction of projections, design, compilation, drafting, and reproduction

7. **Centerline**
   1. A line digitized along the center of a linear feature
   2. A line drawn from the center point of a vertical aerial photograph through the transposed center point of an overlapping aerial photograph

8. **Coordinate system**   A reference system for the unique definition of a location of a point in n-dimensional space

9. **Coordinate reference notation**   Grid coordinates are given in terms of linear measurement (in meters). Geographic coordinates are given in terms of angular measurement (usually in degrees, minutes, and seconds but occasionally in grads).

10. **Coverage**   1. An object in a spatial data base; the representation of a map composed of graphic and attribute files in a digital mapping system.
2. In remote sensing, this term is often used to describe the extent of the earth's surface represented on an image or a set of images

11. **Data model**   Collection of concepts allowing for the representation of an environment according to arbitrary requirements

12. **Datum**   1. Any point, line, or surface used as a reference for a measurement of another quantity.
2. A model of the earth used for geodetic calculations

Explanation of a geodatic datum

The earth surface is highly irregular. Hence it is quite difficult for the geodesist to apply any regular geometric computation for calculation of geodetic parameters e.g. position and height of earth objects etc. To overcome this problem, geodesists have adopted a smooth mathematical surface which approximates the irregular shape of the earth surface to a regular geometric shape. This surface is called the reference surface; it approximates the global mean sea level and is also called the geoid. The parameters required to define such a surface is known as the geodetic datum or datum. For low accuracy computation one can adopt the sphere, but for high accuracy positioning and geodetic computations, a mathematical surface known as ellipsoid has been defined. The ellipsoid is a bi-axial reference surface obtained by revolving the ellipse around its minor axis. Like a geoid reference, its origin, axis and the flattening as given below define the ellipsoid.

Semi major axis = a
Semi minor axis = b
Flattening $f = 1 - (b/a)$
The origin (x, y, z) and the orientation of the ellipsoid with respect to the minor axis $(r, \theta$ and $\varphi)$

Thus, an approximate reference surface which approximates the earth surface and is used for geodetic computation is known as geodetic datum. In other words, a geodetic datum is a mathematical surface, or a reference ellipsoid, with a well-defined origin, axis and orientation of the axis. A geo-centric geodetic datum is a geodetic datum with its origin coinciding with the mass center of the earth, its minor axis coinciding with the axis of revolution of earth and the major axis in the equatorial plane perpendicular to the axis of revolution. This datum is known as earth centered—earth fixed (ECEF).

Therefore, a geodetic datum is uniquely defined by eight parameters viz. two parameters defining the dimension of the datum surface, three parameters defining the position of the origin, and three parameters to define the orientation of the axis with respect to the axis of earth.

Sometimes geodesists use vertical and horizontal datum. The vertical datum surface is a reference surface to which the height (elevation point) of earth objects are referred. Thus the height of a point located on the vertical datum surface is zero. The geoid datum, which approximates the mean sea level, is known as vertical datum. The geoid is commonly known as a surface of zero height.

The world geodetic system is a geo-centric system that provides a basic reference frame and geometric figure for the earth, models the earth gravimetrically, and provides the means for relating positions on various datum to an earth centered?earth fixed coordinate system, the grid reference system

13. **DTED**   (digital terrain elevation data) A geographic matrix of terrain elevation data points converted into a numerical format for computer storage and analysis at precise increments of latitude and longitude. The increment depends upon the latitude and the level of DTED

14. **Ellipsoid**   An ellipsoid is a mathematical figure generated by the revolution of an ellipse about one of its axes. The ellipsoid that approximates the geoid is an ellipse rotated about its minor axis, or an oblate spheroid.

15. **Feature**   An object in a spatial database with a distinct set of characteristics

16. **Geocoding**   1. Synonym for 'address coding'.
                     2. Process of assigning geographic locations to objects.

17. **Geodetic reference system**   The true technical name for a datum. The combination of an ellipsoid, which specifies the size and shape of the earth, and a base point from which the latitude and longitude of all other points are referenced

18. **Geodesy**   Also called geodetics, a branch of earth sciences; the scientific discipline that deals with the measurement and representation of the earth, including its gravity field, in a three-dimensional time-varying space. Besides the earth's gravity field, geodesists also study geodynamical phenomena such as crustal motion, tides, and polar motion. For this they design global and national control networks, using space and terrestrial techniques while relying on datums and coordinate systems.

19. **Geographic database**   A collection of spatial data and related descriptive data organized for efficient storage and retrieval by many users (GIS Development, n.d.).

20. **Georeference**   To establish the relationship between page coordinates on a planar map and known real-world coordinates

21. **'Heads-up' digitizing**   A digitizing station that provides a graphical user interface on the screen of a work station (hence it is sometimes called on-screen digitizing). The operator uses a pointing device (mouse or trackball) to navigate on the scanned image of the original source without having to look down at a digitizing tablet

22. **Geoid**   The equipotential surface in the gravity field of the earth which approximates the undisturbed mean sea level extended continuously through the continents. The direction of gravity is perpendicular to the geoid at every point. The geoid is the surface of reference for astronomic observations and for geodetic leveling.

23. **Grid**   Two sets of parallel lines intersecting at right angles and forming squares; a rectangular Cartesian coordinate system that is superimposed on maps,

charts, and other similar representations of the earth surface in an accurate and consistent manner to permit identification of ground locations with respect to other locations and the computation of direction and distance to other points. Different variants of grids used in different domains are:

a. Major grid/primary grid The primary grid or grids on a map or chart

b. Military grid reference system (MGRS) The alpha-numeric position reporting system used by the US military. A full description is provided in Chapter 3.

c. Non-standard grids Grids other than UTM and UPS, such as the Ceylon Belt, India Zone IIA, West Malaysian RSO (Metric) Grid, etc

d. Operational grid A grid in current operational use—generally this would be the preferred grid but it could be a previously prescribed grid.

e. Overlapping grid A major grid from a neighboring area primarily intended to facilitate military surveying anti-fire control.

f. Preferred grid The grid designated by the DOD for production of new maps, charts, and digital geographic data; and shown on the 'index to preferred grids, datums, and ellipsoids specified for new mapping'

g. Prescribed grid The grid that is locally prescribed by the country of origin or military commander

h. Secondary grid Any grid, other than the primary grid, required for combined operations application. Tick marks along the neat lines are the preferred method of portrayal. Such grids should remain on the maps or charts as long as the secondary grid remains in use.

i. Standard grid   The Universal Transverse Mercator (UTM) grid and the Universal Polar Stereographic (UPS) grid

24. **World geographic reference system (GEOREF)** A world-wide position reference system that may be applied to any map or chart graduated in latitude and longitude regardless of projection. It provides a method of expressing positions in a form suitable for reporting and plotting. The primary use is for inter-service and inter-allied reporting of aircraft and air target positions.

25. **Easting** Eastward (that is left to right) reading of grid values on a map.

26. **Northing** Northward (that is from bottom to top) reading of grid values on a map

27. **False Easting** A value assigned to the origin of eastings, in a grid coordinate system, to avoid the convenience of using negative coordinates.

28. **False Northing** A value assigned to the origin of northings, in a grid coordinate system, to avoid the inconvenience of using negative coordinates.

29.	**Graticule**	A network of lines representing parallels of latitude and meridians of longitude forming a map projection
30.	**Isogonic line**	A line drawn on a map or chart joining points of equal magnetic declination for a given time; the line connecting points of zero declination is the agonic line; lines connecting points of equal annual change are isopors. The Magnetic Variation Chart for the current 5-year epoch is available from the DMACSC.
31.	**Loxodrome**	A line on the surface of the earth cutting all meridians at the same angle, a rhumb line
32.	**Map projection**	An orderly system of lines on a plane representing a corresponding system of imaginary lines on an adopted terrestrial datum surface—a map projection may be derived by geometrical construction or by mathematical analysis.
33.	**Neat line**	The lines that bound the body of a map, usually parallels and meridians (but may be conventional or arbitrary grid lines); also called sheet lines.
34.	**Spheroid**	A mathematical figure closely approaching the geoid in form and size
35.	**Mean sea level (MSL)**	The average height of the surface of the sea for all stages of the tide, used as a reference for elevation measurement.
36.	**Great circle**	A section of a sphere that contains a diameter of the sphere; Sections of the sphere that do not contain a diameter are called small circles.
37.	**Map projection**	A method of representing the earth's three-dimensional surface as a flat two-dimensional surface. This normally involves a mathematical model that transforms the locations of the features on the earth's surface to locations on a two-dimensional surface. Because the earth is three-dimensional, some method must be used to depict the map in two dimensions. Such representations distort some parameter of the earth's surface, be it distance, area, shape, or direction"
38.	**Metadata**	Data about data and its usage aspects
39.	**Map Board**	A mosaic of maps representing the area under military operation
40.	**Orthodrome**	The shortest path between two points on a sphere.
41.	**Open GIS Consortium (OGC)**	The open GIS consortium is a voluntary, non-governmental, non-profit organization dedicated to the development of an open systems approach to geo-processing. The OGC was initiated in August 1994 by the Open Systems Foundation (OSF) to spearhead the Open GeoData Interoperability Specification (OGIS) project
42.	**Open systems**	An information processing system that complies with the requirements of open systems interconnection (OSI) standards in communication with other such systems

43. **Open systems inter-connection (OSI)**   This defines the accepted international standards by which open systems should communicate with each other. It takes the form of a seven-layer model of network architecture, with each layer performing a different function

44. **Orthophoto-graph**   A photographic copy of a perspective photograph with the distortions due to tilt and relief removed

45. **Photo-grammetry**   The art and science of obtaining reliable measurements through use of photographs

46. **Raster**   1. An element of a space that has been sub-divided into regular tiles by tessellation
    2. Commonly, a data set, as for an image or DEM, composed of raster. Often used as a synonym for grid

47. **Remote sensing**   1. Acquisition of information about the properties of an object or phenomenon using a recording device that is not in physical contact with the object of the study.
    2. Commonly, information gathered using air-borne or satellite platforms

48. **Resolution**   A measure of the ability to detect quantities. High resolution implies a high degree of discrimination but has no implication as to accuracy

49. **Shapefile**   An ArcView GIS data set used to represent a set of geographic features such as streets, hospital locations, trade areas, and ZIP code boundaries. Shapefiles can represent points, lines, or area features. Each feature in a shapefile represents a single geographic feature and its attributes (Environmental Systems Research Institute, Inc. [ESRI], 1998).

50. **Spatial data**   Any information about the location and shape of, and relationships among geographic features; this includes remotely sensed data as well as map data

51. **Vector**   1. A quantity which has magnitude and direction.
    2. Commonly, the notation used to represent spatial information

52. **WGS (world geodetic system)**   A geodetic reference system consisting of a set of parameters describing the size and shape of the earth, an earth centered—earth fixed (ECEF) coordinate reference system, the position of a network of points with respect to the center of mass of the earth, and the gravitational model of the earth known as the global geoid.

Appendix C

Geo-Spatial Meta-data

1. Geo-spatial metadata

Scanned topo-sheet	DTED data	Digital vector data	GIS data	Satellite imagery
Topo-sheet no.	Projection system	Vector format (DGN)	DGN file name	Satellite scene number
Scale	Origin coordinates	Map no.	Oracle dump file	Topo-sheets covered
Important areas covered	Extents of the area in terms of lat. and long.	Important areas covered	Topo-sheet no.	Landmarks /key locations covered
Latitude, longitude extents	Easting and northing extents	Easting and northing extents	Easting and northing extents	Easting and northing extents
Publishing date	Source of terrain data from topo-sheet Stereo satellite imagery	Scale of map	Important areas covered	Sensor name
Projection system	Datum used	Projection system longitude extents	Latitude,	Date of imaging
Datum used	Scale of map/ contour interval	Publishing date	Publishing date	Processed image or raw image
Index map Administrative (adjacent boundaries) data Topo-sheet number	Important areas covered	Datum used	Projection system	

(continues...

Table continued)

Scanned topo-sheet	DTED data	Digital vector data	GIS data	Satellite imagery
Topo-sheet name	Important landmarks	Index map Administrative (adjacent boundaries) data Topo-sheet number	Datum used	Type of data pan/ lissindigenous/ ikonos/fused
Edition	Maximum and minimum elevation	Surveyed date	Index map Administrative (adjacent boundaries) data Topo-sheet no.	Resolution (Spatial resolution, Spectral resolution, Radiometric resolution, temporal resolution)
Surveyed date	Pertaining to own area	Contour interval	Topo-sheet name	Date of survey or scan
Contour interval	Pertaining to enemy area	Important land marks	Edition	
Important land marks	Pertaining to own and enemy area	Pertaining to own area	Surveyed date	
Pertaining to own area	International Border	Pertaining to enemy area	Important land marks	
Near Line of Control/ International Border		Near International Border	Pertaining to enemy area	
Secondary zone		Secondary zone	Near International Border	
Storage format (GIF/ TIFF / JPEG, ISFF etc.,)	DTED (0/1/2), DEM	DGN/ DXF/VPF/ Shape	Oracle dump, flat records	Format of image (raw, tiff, geotiff, (RAW, TIFF, geoTIFF, BIL (Band interleaved by line) BIP (band interleaved by Pixel)
Map scale	Area covered $W \times H$	Size of memory	1 cm on paper = x km on ground	Spatial resolution in scale or coverage of the IFOV

2. Facts about Earth

Mass of earth = 5.976×10^{24} kg
The earth's average orbital distance from the sun is 1.496×10^{11} m
Solar day = 24 hrs = 86,400
Sidereal day = 23 hours, 56 minutes, and 4.1 seconds = 86164.10 seconds
Equatorial radius a of the earth often simply referred to as 'the' earth radius is 6378.137 km = 3963.19 miles
Equatorial circumference $(2 \times \pi \times a)$ = 40,075 km = 24,901.5 miles
The flattening of the earth is $f = 0.00335281$, where c is the polar radius $c = a(1 - f)$
The polar radius $b = a\,(1-f)$
Total area of the earth = 196,951,000 sq miles
Total land area = 57,259,000 sq miles (30%)
Total water surface = 139,692,000 sq miles (70%)
Height of highest peak (Mt. Everest) = 8848 meter
Deepest point in Sea: Mariana Trench (Pacific ocean) at 35,840 feet/10,924 meters;
              Java Trench (Indian Ocean) at 23,376 ft/7125 meters

WGS-84 equatorial radius (a) = 6378137.0 meter

WGS-84 flattening (f) = 1/298.257223563

1 nautical mile (nm) = 6076 feet; 1857 meter (international value)

Measurements

1 nautical mile = 6076 feet; 1857 meter (international value)
The earth mass is 5.975×10^{24} kg
The first approximation of the earth is an oblate spheroid
The approximation as a spheroid is accurate enough for most purposes
The equatorial radius a of the earth (often simply referred to as 'the' earth radius) is 6378.137 km (3963.19 statute miles)
The equatorial circumference $(2 \times \pi \times a)$ of earth = 40,075 km
                  = 24,901.5 miles
The flattening of the earth is $f = 0.00335281$, where b is the polar radius and $b = a\,(1-f)$

3. Projection system used by various armed forces and agencies

Name of services/ agencies	Name of projection	Purpose and scale of map
Army	Lambert conformal conic (primary), geographic (secondary)	Operation, intelligence and ops logistics planning, 1:50000, 1:250,000
Navy	Universal Transverse Mercator (UTM) (primary), geographic (secondary)	For sea navigation chart, for sea exercises and sea bed mapping

(continues...

Table continued)

Name of services/ agencies	Name of projection	Purpose and scale of map
Air force	Gnomic cylindrical projection	For flight navigation and mission planning and marking of target
Space and planet exploration, remote sensing organizations	Lambert conformal conic Oblique stereographic projection in spherical form	For landing space vehicles in a precise spot on the planet surface of exploration
National cadastral or land mapping agencies	Albers equal area projection Lambert azimuthal equal-area	For management of land resources and planning of national resources
Geological survey organizations	Bipolar oblique conic conformal projection	Preparation of geologic maps

4. List of important geodetic datum

List of some important geodetic datum are enlisted here for usage in data conversion and so that the reader can get an idea as to the shape of the earth in different places and by different observations

WGS-84 equatorial radius $(a) = 6378137.0$

WGS-84 flattening $(f) = 1/298.257223563$

Datum	Ellipsoid	δa	$\delta f (\times 10^4)$	δx	δy	δz
Adindan	Clarke_1880	−112.145	−0.54750714	−166	−15	+204
Afgooye	Krassovsky	−108.0	0.00480795	−43	−163	+45
Ain_El_Abd_ 1970	International	−251.0	0.14192702	−150	−251	−2
Alaska/Canada _NAD-27	Clarke_1866	−69.4	0.37264639	−9	+151	+185

Datum, Ellipsoid, δa, $\delta f (\times 10^4)$, δx, δy, δz.

δ parameters are with respect to WGS-84 parameters for conversion from the specified datum to WGS-84.

Parameter δa is the WGS-84 equatorial radius minus the specified datum equatorial radius in meters.

Parameter δf is the WGS-84 flattening minus the specified datum flattening multiplied by 10^4.

$\delta x, y, z$ parameters are WGS-84 x, y, z parameters minus the specified datum x, y, z in meters.

The δ *x, y, z* parameters are added to the specified datum *x, y, z* to convert to WGS-84. The source for most of these parameters is the Defense Mapping Agency Technical Report, Department of Defense World Geodetic System 1984, DMA TR 8350.2 Second Edition, 1 September 1991.

WGS-84 Equatorial radius (*a*) = 6378137.0

WGS-84 Flattening (*f*) = 1/298.257223563

The da and db parameters are changes in semi-major axis and semi-minor axis of the ellipsoid with respect to the WGS 84 observations.

5. Table of map scale vs ground distance

Map Scale	Ground distance corresponding to 0.5 mm on map
1:1250	62.5 cm
1:2500	1.25 m
1:5000	2.5 m
1:10000	5 m
1:24000	12 m
1:50000	25 m
1:250 000	125 m

6. Popular spatial data formats

Vector spatial data	File extension and acronym	Raster file format	File extension and acronym
Microstation drawing file format	DGN	Arc digitized raster graphics	ADRG
Dual independent map encoding	DIME	Band interleaved by line	BIL
Digital line graph	DLG	Band interleaved by pixel	BIP
AutoCAD drawing	DWG	Band sequential	BSQ
AutoCAD drawing exchange format	DXF	Device independent bitmap	DIB
Interactive graphic design software	IGDS graphics	Compressed arc digitized raster	CADRG
Initial graphics exchange standard	IGES	Digital terrain elevation data	DTED
Topographically integrated geographic encoding and referencing	TIGER	Graphics interchange format	GIF

(continues...

Table continued)

Spatial data transfer standard	SDTS	Read as imagine	IMG
Topographical vector profile	TVP	ESRI GRID format	GRID
ESRI ArcView GIS	Shapefile	JPEG file interchange format	JFIF
Vector product format	VPF	Multi-resolution seamless image database	MrSID
Computer graphics metafile	CGM	Tag image file format	TIFF GeoTIFF
UK national transfer format	NTF	Portable network graphics	PNG
MOSS export format	MOSS		

7. Modeling of attribute data in GIS

Attribute type → ↓ Geometry of spatial object	Nominal	Ordinal	Ratio
Point	Category of objects represented by a common point object such as hut, tree, survey trig heights etc.	Set of symbols depicting same category of objects with a hierarchical relationship e.g. military installations, government establishments etc.	Same cartographic symbol with different scale/size depicting the magnitude oft he object e.g. building, bridge etc. Sometimes the objects are represented using bar charts (population) or pi charts (educational level) to show the proportionality
Line	Where links between objects/entities are depicted using linear features e.g. communication lines such as road, rail, track etc.	Different types of links are assigned different colors, line style (dash-dot, solid, dotted) etc.	The thickness or intensity of color represents the proportional magnitude of the phenomena such as flow of rivers, traffic capacity of the roads etc.
Polygon/area	Choropleth maps depicting unique features such as height, temperature, slope etc.	Color coded height or color coded temperature.	The magnitude of the feature is depicted through a pattern such as dot density, hatch density or contour density etc.
Surface / volume	Visualizing the terrain objects in 3D, such as 3D perspective view of terrain.	3D Delaunay triangulation in wire frame model	DEM (digital elevation with shaded relief model) depicting the terrain undulation

Appendix D

Frequently Asked Questions in GIS

1. **What is the difference between digital maps and paper maps?**

   Digital maps can be composed thematically; queried for attributes; measurement (area, distance, height, location, perimeter, volume etc) of features can be done up to great precision. They are easy to edit, update; are portable and inter-operable across different electronic platforms. Projection and other geodesic parameters can be changed without affecting the data. Data generated by other systems e.g. GPS, Campus etc can also be plotted programmatically without affecting the original digital map data.

   Editing is not possible on a paper map. Measurement needs offline computation by a skilled map-reader. But one advantage of a paper map is that transportation is easy—it can be carried anywhere in the field for utilization. They also form a lasting impression on the human cognition system.

2. **What kind of geometric information should be given in the margins of a topographic map?**

   a. **Numerical scale:** this is to appear near the graphic scale, and in the upper margins next to the area of coverage.
   b. **Graphic scale:** these will normally be in kilometers and statute miles, with the addition of meters and yards when the scale of the map requires it. If it is necessary, a scale of nautical miles is added. These scales are to be placed in the center of the lower margin.
   c. **Projection, Spheroid, Geodetic datum, Leveling datum:** notes relating to the basis geodetic data
   d. **Notes Concerning the Grid(s):** information is to be given as to the grids to which lines, ticks and figures refer. The projection, spheroid(s), datum(s) origin and false coordinates of origin will be stated for each grid.
   e. **Instructions on the Use of the Grid:** instructions on the use of the grid reference system should show clearly how to give a standard map reference on the sheet.

f. **Unit of elevation:** the note ELEVATIONS IN METRES or ELEVATIONS IN FEET is to appear in a conspicuous position, normally in the lower margin. Wherever possible a conspicuous colour should be used. The normal and preferred unit of elevation is the meter.

g. **Contour interval:** this is to be shown in the lower margin near the graphic scales. It should be in the form: 'Contour interval ... metres (or feet)'. When necessary the note 'Supplementary contours at ... metres (or feet)' is to be added.

h. **Information on True, Grid and Magnetic North:** each map sheet is to contain the information necessary to determine the true, grid and magnetic bearings of any line within the sheet. This information is to be provided in the form of a diagram with explanatory notes.

3. **What is a coordinate system?**

Coordinate systems as a basic method for geo-referencing are used to locate the position of objects in two or three dimensions into a correct relationship with respect to each other.

4. **What kind of coordinate systems are used in mapping?**

- Coordinate systems are often classified into spatial coordinate systems: e.g. spatial geographic and geo-centric coordinate systems and in plane coordinate systems: e.g. 2D Cartesian and polar coordinate systems.
- Generally two types of coordinate systems are given on maps: Cartesian coordinates (or x,y map projection coordinates) and projected geographic coordinates.
- Satellite positioning systems (e.g. GPS) make use of 3-dimensional spatial coordinate systems to define positions on the earth surface with reference to a mean reference surface for the earth (e.g. GPS measurements use the WGS84 ellipsoid).
- 2D Polar coordinates are often used in land surveying. For some types of surveying instruments it is advantageous to make use of this coordinate system.

5. **What is a graticule?**

The graticule represents the projected position of selected meridians (lines with constant longitude λ) and parallels (lines with constant latitude φ).

6. **What is a grid?**

The grid represents lines having constant x or y coordinates and situated at constant intervals depending on map scale.

7. **Why do we need a reference surface?**

The physical surface of the Earth is a complex shape. In order to represent it on a plane, it is necessary to move from the physical surface to a mathematical one, close to the former.

8. **What kinds of reference surfaces are used in mapping?**

   In mapping different surfaces or earth figures are used. These include a geometric or mathematical reference surface, the ellipsoid or the sphere, for measuring locations, and an equipotential surface called the geoid or vertical datum for measuring heights.

9. **What is a vertical datum?**

   The vertical datum, an approximation of the geoid, is defined as natural reference surface for land surveying. A vertical datum fits the mean sea level surface throughout the area of interest and provides the surface to which height ground control measurements are referred.

10. **What is a sphere and when do you use the sphere to approximate the earth's surface?**

    The surface of earth may be taken mathematically as a sphere instead of ellipsoid for maps at smaller scales. In practice, 1:1 000 000–1:5 000 000 is recommended as the largest scale at which the spherical approximation can be made. A sphere can be derived from the ellipsoid corresponding either to the semi-major or semi-minor axis, or average of both axes or can have equal volume or equal surface as the ellipsoid. The sphere represents a rougher approximation than an ellipsoid but reduces the mathematical difficulty.

11. **What is a geodetic datum?**

    The geodetic (or horizontal) datum is defined by the size and shape of an ellipsoid as well as several known positions on the physical surface at which latitude and longitude measured on that ellipsoid are known to fix the position of the ellipsoid.

12. **Why are there so many ellipsoids and datum defined?**

    An ellipsoid and a datum serve as geometric models of the earth surface. They have been established to fit the earth's surface well over the area of local interest. This is important to minimize distortions on maps. About 15 different reference ellipsoids and many more local datum may be encountered in world mapping. Most commonly used ellipsoids are the International, Krasovsky, Bessel, Clark 1880 and the WGS84 ellipsoid. Local ellipsoids serve as reference only for a local area of the earth's surface. Global ellipsoids (e.g. WGS84) serve as a mean reference for the entire earth surface.

13. **What is WGS84?**

    WGS84 is one of the World Geodetic Systems which provides the basic reference frame and geometric figure for the entire earth surface. WGS84 provides a positional relation of various local geodetic systems to an earth centered—earth fixed coordinate system. GPS measurements use the WGS84 as reference surface for their measurements.

14. **What is a map projection?**

A map projection is any transformation between the curved reference surface of the earth and the flat plane of the map. You can also define a map projection as a mathematical formula by which you can transform geographic coordinates (latitude ϕ and longitude λ angles) into Cartesian projection coordinates (x and y)

x, y map projection = $f (\phi, \lambda)$ forward equation. The inverse equations of a map projection are used to transform Cartesian coordinates into geographic coordinates. An overview of map projections equations is given by J.P Snyder's *Map Projections Used by the U.S. Geological Survey.*

15. **What are the parameters of a map projection?**

Map projection equations contain map projection parameters. The most common parameters are:

R = radius of the sphere;

a = equatorial radius or semi-major axes of the ellipsoid of reference;

b = polar radius or semi-minor axes of the ellipsoid of reference;

e = eccentricity of the ellipsoid; f = flattening of the ellipsoid;

h_0 = scale factor at central meridian; h = relative scale factor along a meridian of longitude;

k_0 = scale factor at standard parallel(s); k = relative scale factor along a parallel of latitude;

λ_0 = central meridian or longitude of origin; φ_0 = latitude of origin;

x_0 = false Easting; y_0 = false Northing.

Projection parameters have a significant role in defining a coordinate system. Refer list of parameters described by J.P. Snyder's *Map Projections Used by the United States*, p.xi–xii.

16. **Why do we need a map projection?**

If you are mapping a significant portion of the earth's surface, it is impossible to project it on a flat piece of paper without scale distortions. Map projections take care that the scale distortions remain within certain limits and the distortion pattern of a map projection determines the property of the projection. Each projection has its own characteristics. For example, a map projection may have the property that all angles are correctly represented (conformal projection property). A map projection is not of major importance for city or street maps which cover a relatively small surface of the earth.

17. **How do we classify map projections?**

There are three map projection classes: cylindrical, conical and azimuthal. Map projections can be sub-divided into three aspects: the polar or normal aspect, which centers the map at one pole of the globe; the equatorial aspect, which centers the map at the equator; and the oblique aspect that centers the map anywhere else. Map projections can have the properties: conformal, equal-area or equidistant. A further descriptor is whether the projection has a secant or tangent projection plane.

18. **Why do map projections use mapping zones?**

Some map projections divide the mapping area into zones in order to keep scale distortions within acceptable limits. For example, sixty longitudinal zones are used for the UTM grid system. All these zones are exactly 6° wide in longitude and 164° extent in latitude. Zoning systems have the disadvantage of working with different coordinate systems. Each zone has its own coordinate system.

19. **How do we match adjacent maps?**

In order to fit two or more separate maps exactly along their edges, a number of parameters must be maintained: (1.) the maps must be constructed with the same projection and projection parameters; (2.) they must be at the same scale; and (3.) they should be based on the same reference datum. This is known as map mosaic.

20. **How to select a suitable map projection?**

The choice of a map projection class depends on the size and shape of the geographical area to be mapped—cylindrical projections for large rectangular areas; conic projections for medium-size triangular areas; azimuthal projections for small-size circular areas. The choice of a map projection property has to be made on the basis of the purpose of map—conformal projections for sea, air and meteorological charts, topographic and large scale maps; equidistant projections for topographic and large scale maps; equal-area projections for historical, population, geological and soil maps.

21. **What kind of map projection information should be mentioned on a map?**

Map projection information should at least include the projection name, reference ellipsoid, and the reference datum. It is placed on the map sheet outside the map frame as marginal information. To define a map coordinate system in GIS, detailed information on the projection parameters are required.

22. **What is the difference between paper nautical charts and electronic navigation charts?**

Electronic chart data can be queried (i.e., vector data is smart data) in a variety of ways, which gives the user much more information than a static paper chart can. The navigator can control the display of the ENC (electronic navigation chart) data, which allows for a customized display that only shows information critical to safe navigation. The navigation system software can continuously monitor the ship's position relative to all of the features contained in the ENC. Using ENC display warnings or sound alarms can be generated to detect hazardous situations.

Similarly, the ENC software can check that planned routes will provide safe passage for the vessel by checking for proximity to dangers and crossing areas with insufficient depth.

The paper nautical chart is the fundamental tool of marine navigation. It has served for hundreds of years to convey information about the marine environment and for voyage planning and monitoring. The navigation features of the paper chart are well documented in various publications.

All chart information is available to mariners in a picture in a hardcopy form. Paper charts contains information regarding way points, night time visibility, ALMANAC data, access to chart notes, etc. Unlike ENC it cannot provide online alarms or warnings.

23. **What is database normalization?   What are the various normal forms?**

Normalization is the process of efficiently organizing data in a database.

The normalization process has two goals: eliminating redundant data (for example, storing the same data in more than one table) and ensuring data dependencies make sense (only storing related data in a table). Both of these are worthy goals as they reduce the amount of space a database consumes and ensure that data is logically stored.

The database community has developed a series of guidelines for ensuring that databases are normalized. These are referred to as normal forms and are numbered from one (the lowest form of normalization, referred to as first normal form or 1NF) through five (fifth normal form or 5NF). In practical applications, you will often see 1NF, 2NF, and 3NF along with the occasional 4NF and 5NF.

Appendix E

Expansions of Acronyms

AD	Air defense
AHP	Analytical hierarchy process
API	Application program interface
BIL	Band interleaved by line
BIP	Band interleaved by pixel
BSQ	Band sequential
CAD	Computer aided drafting
CAM	Computer aided manufacturing
CCW	Counter-clockwise
CG	Computational geometry
CGAL	Computational geometric algorithm library
COM	Component object model
COTS	Commercial off the shelf
DAG	Directed acyclic graph
DATE	Decision aid for technology evaluation techniques
DBMS	Data base management system
DCOM	Distributed component model
DEM	Digital elevation model
DGN	Design vector data format
DGPS	Differential global positioning system
DIKD	Data information knowledge decision
DLG	Digital line graph
DLL	Dynamic linked library
DOQs	Digital ortho-photo quadrangles
DRGs	Digital raster graphics
DTED	Digital terrain elevation data
DTM	Digital terrain model
DVD	Digital vector data
DXF	Digital exchange format
ECEF	Earth centered and earth fixed
ENA	Easting northing and altitude

ENC	Electronic navigation chart
GIF	Geographic information format
GIS	Geographical information system
GML	Geographic markup language
GPS	Global positioning system
GUI	Graphic user interface
HOD	Head on display
HTTP	Hyper text transfer protocol
HTML	Hyper text markup language
J2EE	Java 2 Enterprise Edition
LASER	Light amplification by stimulated emission of radiation
LCC	Lambert conformal conic
LiDAR	LIght detection and ranging
LL	Latitude longitude
LOS	Line of sight
MOTS	Military off the shelf
MSL	Mean sea level
MVAC	Model, view, analyze and control
MVC	Model view and control
NC	Nautical chart
NIMA	National Imaging and Mapping Agency
NMEA	National Mariners Engineers' Association
OOD	Object-oriented design
OS	Operating system
OCTREE	Octagonal tree
RADAR	Radio detection and ranging
RAID	Redundant array of discs
RB tree	Red black tree
RDBMS	Relational data base system
RLOS	Radio line of sight
RNC	Raster navigation chart
SDSS	Spatial decision support system
SONAR	Sound navigation and ranging
SP	Shortest path
SQL	Structured query language
SVG	Scalable vector graphic
TIFF	Tagged information file format
TIN	Triangular irregular network
UAV	Unmanned aerial vehicle
USGS	US Geological Survey
VTS	Vessel traffic systems
WMD	Weapons of mass destruction
XML	Extended markup language

Index

Printed and bound by CPI Group (UK) Ltd, Croydon, CR0 4YY

18/10/2024

01776252-0002